Reinforced Grouted BRICK Masonry

James E. Amrhein
Structural Engineer
Executive Director

Masonry Institute of America
2550 Beverly Boulevard
Los Angeles, CA 90057-1085
Tel: (213)388-0472
Fax: (213)389-7514

Originally published by the
Brick Institute of California

First Edition, May 1961
Second Edition, Re-edited February 1962
Third Edition, February 1964
Fourth Edition, Re-edited August 1966
Fifth Edition, Re-edited October 1969
Sixth Edition, April 1971
Seventh Edition, February 1972
Eight Edition, February 1974
Ninth Edition, October 1976
Tenth Edition, October 1978
Eleventh Edition; Revised October 1981
Twelfth Edition; Revised July 1986
Thirteenth Edition, Revised October 1991
MASONRY INSTITUTE OF AMERICA
Los Angeles, California

© Copyright 1991   All rights reserved
No. 107:3M 10-91   ISBN 0-940116-19-7

Masonry Institute of America

ACKNOWLEDGMENTS

While this handbook contains a considerable amount of original text and drawings it also contains data from many sources, including architects and engineers in both private and in public building practice with Los Angeles City, Los Angeles County, International Conference of Building Officials and California Office of the State Architect. We wish to give each of them credit, and gratefully acknowledge their assistance.

With permission, use has been made of copyrighted information contained in:

Uniform Building Code 1991 Edition ©
Published by International Conference of Building Officials

Masonry Design Manual
Published by Masonry Industry Advancement Committee.

BIA Notes
Published by Brick Institute of America

Reinforced Concrete Masonry Construction Inspector's Handbook;
Informational Guide to Grouting Masonry;
Reinforcing Steel in Masonry;
Reinforced Masonry Engineering Handbook
Published by Masonry Institute of America

Drawings by Thomas Escobar

Typing by Petra Stadtfeld

Typesetting by David Willis

I am also pleased to acknowledge the fine editing by Amy K. Hall and the review of material and contributions by Phillip J. Samblanet, P.E.

Cover photo:
Gerontology Center, a masonry building at the University of Southern California, Los Angeles, CA

Reinforced Grouted BRICK Masonry

INTRODUCTION

This handbook will serve as a guide for those interested in or concerned with the construction and inspection of reinforced grouted brick masonry (RGBM) as developed and practiced in Southern California for over 50 years. It is particularly applicable for those areas which have provisions for earthquake safety in their building codes.

The general principles of design are based on working stress design (WSD), applied to reinforced grouted brick masonry, using allowable stresses contained in building codes.

Competent, conscientious and intelligent inspection during construction is fully as important as the design of structures. The best design for earthquake safety will not insure adequacy of structures if compliance with plans and specifications is not rigid and exact.

In the design of a structure, consideration has to be given to the nature of the building, its use, the materials in its construction, and the tie-in to lateral force resistance.

CONTENTS

	Page
FOREWORD	ii
ACKNOWLEDGEMENTS	iii
INTRODUCTION	iv
BRIEF HISTORY	1
BRICK	2-10
Sizes	2
Nomenclature	3
Table of Over-All Masonry Dimensions	4-5
Types	6-7
Dimensional Tolerances	8
Rate of Absorption	8
Field Test for Rate of Absorption	9
Wetting of Brick	9
SAND	10-14
LIME	14-16
CEMENTS	16-18
WATER	18
STEEL	18
STORAGE OF MATERIALS	18
MORTAR	19-23
GROUT	23-24
LOW LIFT AND HIGH LIFT GROUTING	25-33
REINFORCEMENT	33-34
BRICKLAYING	34-43
BRICK VENEER	43
VENEER	44
ANCHORAGE OF VENEER	44-45
STEELTEX METHOD	46

AQUA K LATH METHOD	46-49
BRACING, SHORING and SCAFFOLDING	49
PROTECTION OF WORK	49
CURING OF FINISHED MASONRY	50
CORING OF WALLS	50
TEST SPECIMENS OF MORTAR AND GROUT	50-51
MORTAR PROPORTIONS	52
SPECIAL SEISMIC REQUIREMENTS	53-54
FIELD TEST SPECIMENS OF MORTAR AND GROUT	54-57
EFFLORESCENCE	57-58
CLEANING	59
PAINTING AND WATERPROOFING	59-61
COMPOSITION OF COATINGS	61-64
PARAPET AND FIRE WALLS	62
CAPPING OF WALLS AND SILLS	63
PLANTERS	64
PLASTERING	65
SCHOOLS, HOSPITALS AND PUBLIC BUILDINGS, TITLE 24 C.C.R.	65-68
EARTHQUAKE CONSIDERATIONS	69-70
FIREPLACE AND CHIMNEY NOTES	71-75
TIMELY TIPS FOR INSPECTION	75-78
LAP SPLICES FOR REINFORCING STEEL	78-84
FLUSH WALL COLUMNS	85
COLUMNS TIES	86-91
DETAILS OF BRICK CONSTRUCTION	92-100
GLOSSARY	101-110
GUIDE SPECIFICATIONS	111-118
CLAY BUILDING BRICK TEXTURES	119-121
INDEX	122-130

BRIEF HISTORY

Reinforcing of brick masonry had its first major application in 1825 with the construction of the Thames Tunnel, which used brick caissons, 30 inches thick, 50 feet in diameter and 70 feet deep. Since then the principle has been continually used.

Widespread publicity was given to testing a reinforced brick beam at the Great Exposition in London in 1851, in which "new cement", now know as "portland cement," was used. Portland cement was so named because when made into mortar or concrete it resembled natural stone from Portland, England.

In 1923, Under-Secretary A. Brebner, Government of India, wrote, "In all, nearly 3,009,000 sq. ft. of reinforced brick masonry have been laid in the past three years." S. Kanamori, C.E., Imperial Japanese Government, reported in **Brick & Clay Record,** in 1930: "There is no question that reinforced brickwork should be used instead of plain brickwork when any tensile stress would be incurred in the structure. We can make them more safe and stronger, saving much cost. Further, I have found that reinforced brickwork is more convenient and economical in building than reinforced concrete and, what is still more important, there is always a very appreciable saving in time."

The idea of using cement-sand grout instead of bonding brick headers to bind brick wythes or tiers together and inserting reinforcing steel in the grout space for tensile and shearing resistance was developed for practical and sound engineering use in Southern California beginning about 1935. Since then thousands of tests have been conducted on full size beams, slabs and walls, from which sound engineering design criteria have been established and incorporated into building codes throughout the United States.

Reinforced Grouted BRICK Masonry

BRICK SIZES

Listed below are standard sizes most commonly manufactured by the brick industry in Southern California. In addition, various manufacturers produce sizes other than these standards.

	Height	Width	Length
STANDARD BUILDING BRICK:	$2\frac{1}{2}''$	$\times\ 3\frac{1}{2}''$	$\times\ 7\frac{1}{2}''$
OVERSIZE BUILDING BRICK:	$3\frac{1}{2}''$	$\times\ 3''$	$\times\ 9\frac{1}{2}''$
MODULAR BUILDING BRICK:	$3\frac{3}{8}''$	$\times\ 3''$	$\times\ 11\frac{1}{2}''$
COMMON BUILDING BRICK:	$2\frac{1}{2}''$	$\times\ 3\frac{1}{8}''$	$\times\ 8\frac{1}{4}''$
STANDARDS FACE BRICK:	$2\frac{1}{4}''$	$\times\ 3\frac{7}{8}''$	$\times\ 8\frac{1}{4}''$
	$2\frac{1}{4}''$	$\times\ 3\frac{5}{8}''$	$\times\ 7\frac{5}{8}''$
MODULAR FACE BRICK:	$3\frac{1}{2}''$	$\times\ 3''$	$\times\ 11\frac{1}{2}''$
OVERSIZE FACE BRICK:	$3\frac{1}{2}''$	$\times\ 3''$	$\times\ 9\frac{1}{2}''$
NORMAN FACE BRICK:	$2\frac{1}{4}''$	$\times\ 3\frac{1}{2}''$	$\times\ 11\frac{1}{2}''$
REGENCY FACE BRICK:	$3\frac{1}{2}''$	$\times\ 3''$	$\times\ 11\frac{1}{2}''$
IMPERIAL FACE BRICK	$5\frac{1}{2}''$	$\times\ 3''$	$\times\ 15\frac{1}{2}''$
5" HOLLOW BRICK:	$3\frac{1}{2}''$	$\times\ 4\frac{1}{2}''$	$\times\ 11\frac{1}{2}''$
IMPERIAL BRICK:	$5\frac{1}{2}''$	$\times\ 3''$	$\times\ 15\frac{1}{2}''$
FOUR-16" BRICK:	$3\frac{1}{2}''$	$\times\ 3''$	$\times\ 15\frac{1}{2}''$
ROYALE BRICK BLOCK:	$3\frac{1}{2}''$	$\times\ 7\frac{5}{8}''$	$\times\ 15\frac{1}{2}''$
	$5\frac{1}{2}''$	$\times\ 7\frac{5}{8}''$	$\times\ 15\frac{1}{2}''$
	$7\frac{1}{2}''$	$\times\ 7\frac{5}{8}''$	$\times\ 15\frac{1}{2}''$
PAVING BRICK:	$1\frac{1}{4}''$	$\times\ 3\frac{3}{4}''$	$\times\ 8''$
	$1\frac{1}{4}''$	$\times\ 3\frac{5}{8}''$	$\times\ 7\frac{5}{8}''$
SPLIT PAVER BRICK:	$1\frac{1}{2}''$	$\times\ 3\frac{7}{8}''$	$\times\ 8\frac{1}{4}''$
BEL AIR PAVER BRICK:	$1\frac{3}{8}''$	$\times\ 5\frac{1}{2}''$	$\times\ 11\frac{1}{2}''$
COMMON PAVER BRICK:	$1\frac{3}{8}''$	$\times\ 3\frac{7}{8}''$	$\times\ 8\frac{1}{4}''$

The Masonry Institute of America offers a set of 9 masonry layout plastic rules comprised of 28 brick scales for vertical and horizontal layout of masonry.

Masonry Institute of America

NOMENCLATURE

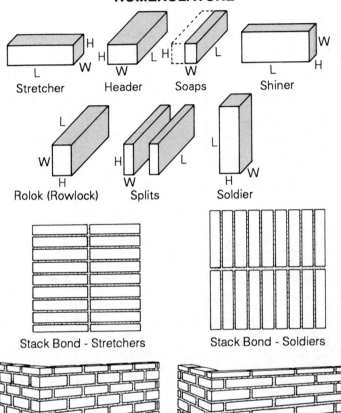

Stretcher   Header   Soaps   Shiner

Rolok (Rowlock)   Splits   Soldier

Stack Bond - Stretchers   Stack Bond - Soldiers

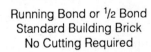

Running Bond or ½ Bond
Standard Building Brick
No Cutting Required

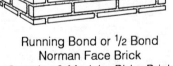

Running Bond or ½ Bond
Norman Face Brick
Oversize & Modular Bldg. Brick
Cut corner brick to fit

¼ Bond, Modular Building Brick
⅓ Bond, Oversize Building Brick
⅓ Bond, Norman Face Brick

Reinforced Grouted BRICK Masonry

Table of Over-All Masonry Dimensions Using Oversize Building Blocks, 3¹/₄″ x 3¹/₄″ x 10″
Horizontal dimensions equal multiples of stretchers plus one header laid with ¹/₃ bond

1/2″ Mortar Joint		5/8″ Mortar Joint		3/4″ Mortar Joint	
Vertical	Horizontal	Vertical	Horizontal	Vertical	Horizontal
3³/₄″	1′ - 1³/₄″	3⁷/₈″	1′ - 1⁷/₈″	4″	1′ - 2″
7¹/₂″	2′ - 0¹/₄″	7³/₄″	2′ - 0¹/₂″	8″	2′ - 0³/₄″
11¹/₄″	2′ - 10³/₄″	11⁵/₈″	2′ - 11¹/₈″	1′ - 0″	2′ - 11¹/₂″
1′ - 3″	3′ - 9¹/₄″	1′ - 3¹/₂″	3′ - 9³/₄″	1′ - 4″	3′ - 10¹/₄″
1′ - 6³/₄″	4′ - 7³/₄″	1′ - 7³/₈″	4′ - 8³/₈″	1′ - 8″	4′ - 9″
1′ - 10¹/₂″	5′ - 6¹/₄″	1′ - 11¹/₄″	5′ - 7″	2′ - 0″	5′ - 7³/₄″
2′ - 2¹/₄″	6′ - 4³/₄″	2′ - 3¹/₈″	6′ - 5⁵/₈″	2′ - 4″	6′ - 6¹/₂″
2′ - 6″	7′ - 3¹/₄″	2′ - 7″	7′ - 4¹/₄″	2′ - 8″	7′ - 5¹/₄″
2′ - 9³/₄″	8′ - 1³/₄″	2′ - 10⁷/₈″	8′ - 2⁷/₈″	3′ - 0″	8′ - 4″
3′ - 1¹/₂″	9′ - 0¹/₄″	3′ - 2³/₄″	9′ - 1¹/₂″	3′ - 4″	9′ - 2³/₄″
3′ - 5¹/₄″	9′ - 10³/₄″	3′ - 6⁵/₈″	10′ - 0¹/₈″	3′ - 8″	10′ - 1¹/₂″
3′ - 9″	10′ - 9¹/₄″	3′ - 10¹/₂″	10′ - 10³/₄″	4′ - 0″	11′ - 0¹/₄″
4′ - 0³/₄″	11′ - 7³/₄″	4′ - 2³/₈″	11′ - 9³/₈″	4′ - 4″	11′ - 11″
4′ - 4¹/₂″	12′ - 6¹/₄″	4′ - 6¹/₄″	12′ - 8″	4′ - 8″	12′ - 9³/₄″
4′ - 8¹/₄″	13′ - 4³/₄″	4′ - 10¹/₈″	13′ - 6⁵/₈″	5′ - 0″	13′ - 8¹/₂″
5′ - 0″	14′ - 3¹/₄″	5′ - 2″	14′ - 5¹/₄″	5′ - 4″	14′ - 7¹/₄″
5′ - 3³/₄″	15′ - 1³/₄″	5′ - 5⁷/₈″	15′ - 3⁷/₈″	5′ - 8″	15′ - 6″
5′ - 7¹/₂″	16′ - 0¹/₄″	5′ - 9³/₄″	16′ - 2¹/₂″	6′ - 0″	16′ - 4³/₄″
5′ - 11¹/₄″	16′ - 10³/₄″	6′ - 1¹/₈″	17′ - 1¹/₈″	6′ - 4″	17′ - 3¹/₂″
6′ - 3″	17′ - 9¹/₄″	6′ - 5¹/₂″	17′ - 11³/₄″	6′ - 8″	18′ - 2¹/₄″
6′ - 6³/₄″	18′ - 7³/₄″	6′ - 9³/₈″	18′ - 10³/₈″	7′ - 0″	19′ - 1″
6′ - 10¹/₂″	19′ - 6¹/₄″	7′ - 1¹/₄″	19′ - 9″	7′ - 4″	19′ - 11³/₄″

Masonry Institute of America

7'- 2 1/4"	20'- 4 3/4"	7'- 5 5/8"	20'- 7 5/8"	7'- 8"	20'- 10 1/2"		
7'- 6"	21'- 3 1/4"	7'- 9"	21'- 6 1/4"	8'- 0"	21'- 9 1/4"		
7'- 9 3/4"	22'- 1 3/4"	8'- 0 7/8"	22'- 4 7/8"	8'- 4"	22'- 8"		
8'- 1 1/2"	23'- 0 1/4"	8'- 4 3/4"	23'- 3 1/2"	8'- 8"	23'- 6 3/4"		
8'- 5 1/4"	23'- 10 3/4"	8'- 8 5/8"	24'- 2 1/8"	9'- 0"	24'- 5 1/2"		
8'- 9"	24'- 9 1/4"	9'- 0 1/4"	25'- 0 3/4"	9'- 4"	25'- 4 1/4"		
9'- 0 3/4"	25'- 7 3/4"	9'- 4 3/8"	25'- 11 3/8"	9'- 8"	26'- 3"		
9'- 4 1/2"	26'- 6 1/4"	9'- 8 1/4"	26'- 10"	10'- 0"	27'- 12 3/4"		
9'- 8 1/4"	27'- 4 3/4"	10'- 0 5/8"	27'- 8 5/8"	10'- 4"	28'- 0 1/2"		
10'- 0"	28'- 3 1/4"	10'- 4"	28'- 7 1/4"	10'- 8"	28'- 11 1/4"		
10'- 3 3/4"	29'- 1 3/4"	10'- 7 7/8"	29'- 5 7/8"	11'- 0"	29'- 10"		
10'- 7 1/2"	30'- 0 1/4"	10'- 11 3/4"	30'- 4 1/2"	11'- 4"	30'- 8 3/4"		
10'- 11 1/4"	30'- 10 3/4"	11'- 3 5/8"	31'- 3 1/8"	11'- 8"	31'- 7 1/2"		
11'- 3"	31'- 9 1/4"	11'- 7 1/2"	32'- 1 3/4"	12'- 0"	32'- 6 1/4"		
11'- 6 3/4"	32'- 7 3/4"	11'- 11 3/8"	33'- 0 3/8"	12'- 4"	33'- 5"		
11'- 10 1/2"	33'- 6 1/4"	12'- 3 1/4"	33'- 11"	12'- 8"	34'- 3 3/4"		
12'- 2 1/4"	34'- 4 3/4"	12'- 7 1/8"	34'- 9 5/8"	13'- 0"	35'- 2 1/2"		
12'- 6"	35'- 3 1/4"	12'- 11"	35'- 8 1/4"	13'- 4"	36'- 1 1/4"		
12'- 9 3/4"	36'- 1 3/4"	13'- 2 7/8"	36'- 6 7/8"	13'- 8"	37'- 0"		
13'- 1 1/2"	37'- 0 1/4"	13'- 6 3/4"	37'- 5 1/2"	14'- 0"	37'- 10 3/4"		
13'- 5 1/4"	37'- 10 3/4"	13'- 10 5/8"	38'- 4 1/8"	14'- 4"	38'- 9 1/2"		
13'- 9"	38'- 9 1/4"	14'- 2 1/2"	39'- 2 3/4"	14'- 8"	39'- 8 1/4"		
14'- 0 3/4"	39'- 7 3/4"	14'- 6 3/8"	40'- 1 3/8"	15'- 0"	40'- 7"		
14'- 4 1/2"	40'- 6 1/4"	14'- 10 1/4"	41'- 0"	15'- 4"	41'- 5 3/4"		
14'- 8 1/4"	41'- 4 3/4"	15'- 2 1/8"	41'- 10 5/8"	15'- 8"	42'- 4 1/2"		
15'- 0"	42'- 3 1/4"	15'- 6"	42'- 9 1/4"	16'- 0"	43'- 3 1/4"		
15'- 3 3/4"	43'- 1 3/4"	15'- 9 7/8"	43'- 7 7/8"	16'- 4"	44'- 2"		
15'- 7 1/2"	44'- 0 1/4"	16'- 1 3/4"	44'- 6 1/2"	16'- 8"	45'- 0 3/4"		

BRICK

There are two types of building brick in Southern California: **Building Brick** and **Face Brick**. Building bricks are made in shades of red of the sizes shown herein with smooth or textured faces, and are made of clay just as it comes from the pit or banks.

Face brick is made from controlled mixtures of clays or shales containing certain minerals which, when fired in a kiln, produce various shades of color. No artificial colors are added.

The physical difference between face brick and building brick is that the former are required to meet strict tolerances for color, size, warpage and chippage.

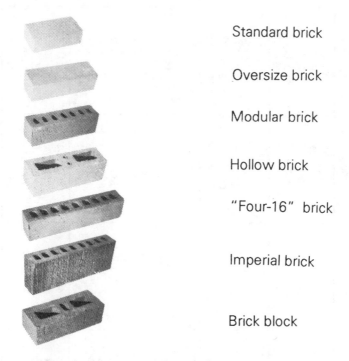

Standard brick

Oversize brick

Modular brick

Hollow brick

"Four-16" brick

Imperial brick

Brick block

Figure 1. Building (common) brick, from top. Standard size, available smooth or textured: Oversize, available smooth, wire cut or textured: Modular size, wire cut, smooth or textured: Four-16, available smooth or textured: Imperial, smooth or textured: Royal brick block, available smooth and textured faces and in various sizes.

Masonry Institute of America

Standard face brick

Norman face brick

Figure 2. Face brick, from top, standard size, smooth or textured: Norman size, smooth or textured face: Face bricks are available in many different colors.

All local building codes including Title 24 of the California Code of Regulation (C.C.R.), and most private specifications require brick to meet ASTM Specifications C 62 and C 216 for building brick and face brick, respectively. All brick in Southern California are manufactured to meet these respective specifications.

As stated in ASTM C 62: "The brick, as delivered to the site, must, by visual inspection, conform to the requirements specified by the purchaser or to the sample or samples approved as the standard of comparison and to the samples passing the tests for physical requirements. Minor indentations or surface cracks incidental to the usual method of manufacture, or the chipping resulting from the customary methods of handling in shipment and delivery, should not be deemed grounds for rejection.

"The brick shall be free of defects, deficiencies, and surface treatments, including coatings, that would interfere with the proper setting of the brick or significantly impair the strength or performance of the construction.

"Unless otherwise agreed upon by the purchaser and the seller, a delivery of brick may contain not more than 5% broken brick."

Dimensional Tolerances

Building brick, ASTM C 62 lists the following permissible variations in dimensions:

Specified dimension, in.	Maximum Permissible Variations from Specified Dimension, plus or minus, in.
Up to 3 incl.	3/32
Over 3 to 4 incl.	1/8
Over 4 to 6 incl.	3/16
Over 6 to 8 incl.	1/4
Over 8 to 12 incl.	5/16
Over 12 to 16 incl.	3/8

Face brick, ASTM C 216 lists the following tolerances on dimensions.

Specified Dimension, in.	Maximum Permissible Variations from Specified Dimension, Plus or Minus, in.	
	Type FBX	Type FBS
3 and under	1/16	3/32
Over 3 - 4 incl.	3/32	2/16
Over 4 - 6 incl.	2/16	3/16
Over 6 - 8 incl.	5/32	4/16
Over 8 - 12 incl.	7/32	5/16
Over 12 - 16 incl.	9/32	3/8

Rate of Absorption, ASTM C 67

The initial rate of absorption of a brick has an important effect on the bond between brick and mortar. Maximum bond strength occurs when the suction of the brick at the time of laying is between 5 and 20 gm. per minute per 30 sq.in. of brick surface immersed in 1/8″ of water. Brick having a higher absorption should be moistened to reduce suction. Figure 3 shows typical conditions.

There is no consistent relationship between **total** absorption and **rate** of absorption. Some bricks have high total absorption and vice versa.

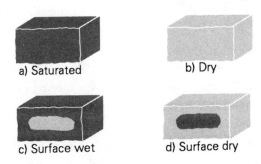

Figure 3. Moisture conditions of inside of brick.

Figure 3a shows a saturated condition where the brick has so low a suction that it will float unless the bricklaying proceeds very slowly. Figure 3b shows a dry brick. Figure 3c shows a wet surface but dry core condition, and is satisfactory if the moisture extends at least $3/4''$ into the brick. The reason for the $3/4''$ of moisture is that in warm weather the brick may otherwise dry out. Figure 3d shows the ideal condition where the core is wet and the surface is damp. One method to achieve this condition is to saturate the brick the day before use.

Field Test for the Rate of Absorption

A rough but effective test for a rate of absorption is to place a quarter on a brick draw a circle around it with a pencil, then trace around the circle using a wax crayon. Using a medicine dropper, very quickly drop water within the circle, completely filling it, taking care that the water does not flow over the marked circle, and continue until 20 drops have been placed. Note the time required for all of the water to be absorbed into the brick, beginning with the time the circle is first filled. If the time exceeds $1\frac{1}{2}$ min., the brick need not be wetted; if less than $1\frac{1}{2}$ min., they must be pre-wetted.

Wetting of Brick

A poor bond between brick and mortar may result from either too much or too little water in both mortar and brick.

Brick must be wet with clear water. Several methods can be used in wetting brick. One is letting a hose run on the pile or pallets of brick;

another is using a sprinkler that would wet a number of pallets at one time, or they may be hand sprayed by holding the hose.

Bricks should be wetted several hours before they are to be laid. This allows the moisture to penetrate into the interior of the brick. The surfaces should be just damp when laid. The time of wetting and the amount of wetting will depend on weather conditions.

Some face bricks having a very low rate of absorption and need not be wetted, while others need be only dampened before laying. When bricks are thoroughly soaked, as after a hard rain, they must be allowed to dry until the initial rate of absorption is restored.

SAND, ASTM C 144

Quality. Sand for mortar should consist of hard, strong, durable uncoated mineral or rock particles free from mica, organic matters, saline, alkaline, or other deleterious substances which would impair the strength of the mortar or which would cause disintegration or efflorescence.

Ocean or beach sand should never be used, as it is not properly graded and since it contains deleterious matter and salt, which contribute to efflorescence. Ocean or beach sand may also contain soap-like elements which will cause disintegration.

Sand for mortar which may be either natural or manufactured, is the largest volume and weight constituent of the mortar. Sand acts as an inert filler, providing economy, workability and reduced shrinkage, while influencing compressive strength. An increase in sand content increases the setting time of a masonry mortar, but reduces potential cracking due to shrinkage of the mortar joint.

Well graded sand reduces separation of materials in plastic mortar, which reduces bleeding and improves workability. Sands deficient in fines produce harsh mortars, while sands with excessive fines produce weak mortars and increase shrinkage. High lime or high air content mortars can hold more sand, even with poorly graded aggregates, and still provide adequate workability.

Field sands deficient in fines can result in the cementitious material acting as fines. Excess of fines in the sand, however, is more common and can result in oversanding, since workability is not substantially affected by such excess.

Mortar properties are not seriously affected by small variations in grading, but quality is improved by careful attention to aggregate selection.

Grading. Aggregate for masonry mortar must be graded within the following limits set by ASTM C 144 whether sand is natural or manufactured.

Sieve Size	Percent Passing[1]	
	Natural Sand	Manufactured Sand
No. 4	100	100
No. 8	95 to 100	95 to 100
No. 16	70 to 100	70 to 100
No. 30	40 to 75	40 to 75
No. 50	10 to 35	20 to 40
No. 100	2 to 15	10 to 25
No. 200	...	0 to 10

1. ASTM C144.

Fineness Modulus. A fineness modulus of mortar sand between the approximate limits of 2.25 and 2.75 is recommended. Fineness modulus is the summation of the percentages of dry sand retained on standard sieve numbers 4, 8, 16, 30, 50 and 100 divided by 100.

Figure 4. Nest of sand analysis sieves in shaking apparatus. Knocker at top taps nest while being shaken.

Reinforced Grouted BRICK Masonry

Voids. Voids in sand may be defined as the empty spaces among the grains of sand. If a metal cylinder is filled with sand, an amount of water may be poured into it without overflowing, since the water simply fills up the air spaces. When the sand has settled down, and is saturated without free water standing on top, the volume of water required to saturate the sand would represent the volume of voids in the sand. In commercially graded sand, this void volume is closely $1/3$ the volume of loose sand. Thus the accepted rule is 1 volume of cementitious materials to 3 volumes of sand.

A material having particles all of a uniform size and shape contains practically the same percentage of voids as a material having particles of a corresponding similar shape, but of a different uniform size.

In any loose or granular material, the largest percentage of voids occurs with particles all of the same size. The smallest percentage of voids occurs with particles of such different sizes that the voids of each size are filled with the largest particles which will fill them.

Materials with round particles contain less voids than materials with angular particles screened to the same size.

Figure 5. Sieves showing amounts of sand retained on (top, left to right) No's. 8, 16 and 30 sieves, and (2nd row) No's, 50, 100 and 200 sieves. Bottom pan, amount passing No. 200 sieve. This sand has a fineness modulus of 2.40.

1 cubic foot contains	1728 cubic inches
12" diam. sphere contains	904.8 cubic inches
1728 1" diam. stacked sphere contains	904.8 cubic inches
1 cubic foot of stacked buck shot contains	904.8 cubic inches
In each case	47.7% voids
Graded sand about	34% voids

Requirements. Most local building codes and Title 24 C.C.R. require sand to meet ASTM Specifications C 144 which provides a definite sand gradation and a limit of 2 to 15% on fine materials passing the No. 100 sieve. Additionally Title 24 C.C.R. requires a minimum of 3% sand passing the No. 100 sieve in order to provide more fine particles which produce a more workable and compact mass.

In washed sand, fines are washed out, producing a harsher mortar. To compensate for the lack of fines, the mortar requires more lime or cement to provide the required plasticity and water retention.

Bulking. Where the volume of sand is specified on dry, loose measurement, and the sand on the job is damp, it is necessary to allow for bulking.

A simple method of determining this is to carefully fill a 6" × 12" cylinder level full of the damp sand, without tamping, then carefully dry out this sand, after which, place it back in the cylinder. The distance the surface of the dry sand is below the top of the cylinder, divided by the depth of the sand, will give the percentage of damp sand to be added to obtain the specified volume. If this distance is, say 2", then (2 divided by 10) 20% bulking must be added to the sand volume.

Another method is to pour enough water into a sand-filled cylinder until the sand is completely saturated and settled and the cylinder is level full of water. The distance from the top of the cylinder to the top of the saturated sand is closely the same as in the preceding paragraph.

Reinforced Grouted BRICK Masonry

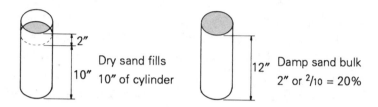

Figure 6. Bulking of damp sand.

LIME, ASTM C 5, C 207

Of all the materials in the mortar mix, lime has the greatest capacity to retain water. This is a very important function of the lime, because it aids workability, and is also a cementitious material. Lime is required in masonry mortar by most local building codes, Uniform Building Code, Title 24 C.C.R., and in ASTM and Federal Specifications.

Figure 7. LIME—From left, lump quick-lime, processed quick-lime, hydrated lime.

CYCLE: Limestone to Lime to Limestone
It may be Calcium Carbonate or Magnesium Carbonate, or both.

EXAMPLE:

	$CaCO_3$	= Calcium Carbonate (Limestone)
Minus,	CO_2	= Carbon Dioxide (Lost in burning Limestone)
Result,	CaO	= Carbon Oxide (Quicklime)
Add,	H_2O	= Water to slack Quicklime
Result,	$Ca(HO)_2$	= Calcium Hydroxide (lime Putty)
Plus,	CO_2	= Carbon Dioxide from air
Result,	$CaCO_3$	= Calcium Carbonate (Limestone)

Quicklime in lumps, called **lump lime,** is slacked by adding lime to the water.

Processed lime (pulverized quicklime) is slacked by adding the lime to the water.

Hydrated lime is the same, chemically, as lime putty, but only enough water is added to quicklime in manufacture to satisfy its chemical demand. Water added to hydrated lime makes lime putty in a very few minutes.

Lime putty is made from quicklime or hydrated lime. If made from quicklime, other than pulverized (processed) quicklime, the lime must be properly slacked and then screened through a sieve having not less than 16 meshes per linear inch. After screening and before using, the slacked lime must be properly stored and protected for not less than 10 days or longer until entirely cooled. If made of pulverized (processed) quicklime, the lime must be properly slacked for not less than 24 hours or until the lime putty has entirely cooled. All resulting lime putty should not weigh less than 83 pound per cubic foot.

The crystallization of lime not only increases the bond strength, but, over a period of many years, heals or seals fine cracks caused by shrinkage or other causes.

Whether lime putty or dry hydrated lime is used, the proportions by volume measurement are the same for either type.

CEMENTS, ASTM C 150

Portland Cement. Portland cement is made by finely pulverizing the clinker which is produced from burning a definite artificial mixture of silicious (containing silica), argillaceous (containing alumina), and calcareous (containing lime) materials to incipient fusion. Small amounts of alkalies, such as potassium oxide, sodium oxide, and sulfur trioxide are also present.

Portland cement is composed of three principal compounds: tri-calcium aluminate, tri-calcium silicate, and di-calcium silicate. As these compounds crystalize, the portland cement sets and hardens as follows:

1. Approximately $2\frac{1}{2}$ hours after mixing, the initial set will occur as the first set of crystals form.

2. Final set occurs after about 8 hours with the formation of the second crystals.

3. Hard set occurs with the formation of the last crystals. Although the portland cement will continue to harden indefinitely, hard set is usually considered to have occurred 4 weeks after mixing.

Portland cement in mortar does not chemically begin to set up until about $2\frac{1}{2}$ hours after mixing and therefore can be retempered with water up to this time without loss in strength. However, some specifications limit the retempering time to one hour after mixing. Tests have been made on mortar in increments of time up to 4 hours and very little loss in compressive strength was noted. ASTM C 270, *Mortar For Unit Masonry*, requires mortar to be retempered with water to maintain workability, and to be used and placed within $3\frac{1}{2}$ hours after the original mixing. This depends on temperature.

Plastic Cement. In some of the southwestern areas of the United States, plastic cement is sometimes used for mortar. This is basically Type I portland cement with approximately 12 percent plasticizing agent added. When plastic portland cement is used in mortar, hydrated lime may be added, but not in excess of one-tenth of the volume of cement.

Plastic cement is generally used for small masonry projects and the "do it yourself" home masonry market since lime does not have to be used to obtain adequate plasticity. Mortar made with 1 part plastic cement and 3 parts sand is equivalent to a mix of 1 part portland cement, 0.14 parts plasticizer and 3.4 parts sand which is richer than type S, portland cement, lime mortar.

Preblended Cement. In some parts of the United States, portland cement manufacturers and some masonry materials supply companies may package a blend of portland cement and hydrated lime. Blending should be according to prepackaged specifications or in proportion by volume for each mortar type, as follows:

- Type M mortar, one part portland cement and $\frac{1}{4}$ part lime
- Type S mortar, one part portland cement and $\frac{1}{2}$ part lime
- Type N mortar, one part portland cement and one part lime.

This packaged mortar cement should conform to the requirements of Table 24-A of the Uniform Building Code. It is generally used for small projects or jobs where separate delivery of portland cement and lime is inconvenient.

Mortar Cement. Mortar cement is also a portland cement based material which meets the requirements of *UBC Standard No. 24-19, Mortar Cement*. Mortar cement may be used for mortar in all seismic zones.

There are three types of mortar cement:

(1) Type N. Contains the cementitious materials used in the preparation of UBC Standard No. 24-20 Type N or Type O mortars. Type N mortar cement may also be used in combination with portland or blended hydraulic cements to prepare Type S or Type M mortars.

(2) Type S. Contains the cementitious materials used in the preparation of UBC Standard No. 24-20 Type S mortar.

(3) Type M. Contains the cementitious materials used in the preparation of UBC Standard No. 24-20 Type M mortar.

Masonry Cement. Masonry cement is a mixture of portland cement, 30% to 60% plasticizer material, and other added chemicals. The UBC

Reinforced Grouted BRICK Masonry

Standard No. 24-16, *Cement, Masonry* is based on ASTM C 91.

Both specifications have been recently revised to include Type M, Type S and Type N masonry cements which will produce mortars conforming to ASTM C 270. Additionally, Type N masonry cement may be mixed with additional portland cement to produce Type M and Type S mortar.

WATER

Water used in masonry construction should be potable, suitable for drinking, and free of harmful substances such as oil, acids, alkalis, etc.

STEEL, ASTM A 615

Reinforcing steel is invariably specified to comply with ASTM A 615 which specify physical and chemical properties and deformations for bonds. Laboratory reports are required when steel is specified to be tested.

UBC, Section 2407(e)3(iii)

(iii)Joint reinforcement. Prefabricated joint reinforcement for masonry walls shall have at least one crosswire of at least No, 9 gauge for each 2 square feet of wall area. The vertical spacing of the joint reinforcement shall not exceed 16 inches. The longitudinal wires shall be thoroughly embedded in the bed joint mortar. The joint reinforcement shall engage all wythes.

Where the space between metal tied wythes is solidly filled with grout or mortar the allowable stresses and other provisions for masonry bonded walls shall apply. Where the space is not filled, tied walls shall conform to the allowable stress, lateral support, thickness (excluding cavity), height and tie requirements for cavity walls.

STORAGE OF MATERIALS

Brick units, cement and sacked lime should be protected from exposure to wet weather. Cement and lime may deteriorate when exposed to excessive moisture.

Masonry Institute of America

MORTAR, ASTM C 270

The most important properties of mortar are:
 ... Workability and water retentivity
 ... Capacity to develop bond
 ... Weather resistance
 ... Minimum volume change

The recommended proportions for mortar for RGBM and brick veneer are:

1 part portland cement

½ part hydrated lime

4½ parts mortar sand, damp loose measurement.

All parts must be determined by accurate volume measurements at the time of placing in the mixer. When less than one full sack of cement is used, extreme care must be taken in very accurately measuring all the parts. Title 24 C.C.R. does not permit less than full sack batches unless materials are weighed.

Compressive strength is not a function of the bonding value of mortar.

Full size tests have shown that brick masonry using a mortar mix of 1 part portland cement, ½ part lime, 4½ parts sand has bond value in flexure much higher than when mortar of 1 part cement, ¼ part lime, 3 parts sand when used.

With any mortar mix, the softer or more plastic the mortar, the greater the bond strength. For maximum bond, the mortar must be as soft as the mason can handle it without smearing the face of the brick.

Mortar must be maintained highly plastic on the mortar boards; this usually requires retempering with water. This should be done only by adding water within a basin formed with the mortar and the mortar reworked into water. Water should not be dashed over mortar on the boards. Harsh, non-plastic mortar must not be retempered or used. Since mortar will not set up chemically prior to about 2½ hours, it is amply safe to use it up to the 1 hour limit contained in most specifications.

Water Retention of Mortars

Water retentivity is the property of mortar which resists rapid loss of the mixing water to the air or to overly absorptive masonry units. Water retentivity is often related to workability since mortars with good water retention properties tend to remain soft and plastic for an ample period of time. Therefore, masonry units may be aligned, leveled and plumbed without breaking, the critical bond between the mortar and the masonry units. Conversely rapid loss of water from the mortar causes premature setting of the mortar and poor bond while excessively high water retention mortars (or non-absorptive units) cause the masonry units to float so that it is difficult to properly align them. Thus, mortars should be mixed to have water retention properties within tolerable limits.

Water retention for mortar is determined in a laboratory by measuring the flow on a flow table as outlined in ASTM C 91. Mortar is placed on an apparatus which, by vacuum, sucks water out of the mortar for one minute. The mortar is then returned to the flow table and again measured. If the mortar has lost no water, the flow after suction would be 100%.

In ASTM C 91, Table 1, a 70% water retention value is recommended. This is based on mortar having an original flow of 100 to 115%. Mortar, when used, however, should generally have an original flow of 135 to 145%, and thus would have a higher percent flow after suction. Generally, the water retention of a mortar at the time of use should be about 85%.

Water Retention Determined from ASTM C 91, Section 25.

Figure 8. The "flow" and water retention of mortar is determined as follows: A truncated brass cone is placed on the flow table. Cone is filled with mortar and levelled off.

Cone Dimensions:
Diam. at top $2^3/4''$
Diam. at bottom $4''$
Height of cone $2''$

Masonry Institute of America

Figure 9. Cone is carefully removed.

Figure 10. Table is mechanically "dropped" 25 times in 15 seconds, and "flow" is measured. Cone diameter at base is 4 inches. If mortar flows out to 9.60 inches, the increase in diameter is 5.60 inches, and the "flow" is 5.6 divided by 4, or 140 percent.

Reinforced Grouted BRICK Masonry

Figure 11. Mortar taken from flow table is placed into perforated dish of suction apparatus, and 2 inches of mercury sucks water from mortar for one minute.

Figure 12. Mortar is returned to the flow table and measured for flow. If it then flows to 8.5 inches diameter, the increase is 4.5 inches. The flow after suction is 4.5 divided by 4, or 113% the water retention is the ratio of 113% divided by 140, or 80% which is ideal for good brick mortar.

Admixtures

No admixtures should be used in mortar and grout unless approved by the governing building department, and without the written approval of the architect or engineer. Tests and experience must demonstrate that the admixture will not decrease the bond strength to the brick. Some admixtures reduce the water content or water demand, and thus may adversely affect the water retentivity of mortar.

Mortar Colors

Color additives for mortar may be only pure mineral oxides, carbon black or synthetic colors. Carbon black has a maximum limit of three percent of the weight of the cement in the mortar. Colors for the mortars must be approved by the architect or engineer and measured accurately into each batch after all other ingredients are in the mixer. Then all ingredients should be mixed thoroughly to produce an even color.

GROUT, ASTM C 476

Grout differs from concrete in that concrete is poured with a low water-cement ratio, into non-porous forms, while grout has a high water-cement ratio, and is poured between porous brick forms. Additionally grout must be sufficiently fluid to flow into small grout spaces and around reinforcing steel without leaving voids. Although good mortar should stick to the bottom side of a trowel, it should be impossible for grout to do so. When introduced into the masonry the high water-cement ratio of grout is rapidly reduced to an extremely low ratio effecting a compact mass having a density about equal to that of gunite.

In construction of RGBM, the local codes and Title 24 C.C.R. permit one outer tier of brickwork to be carried up not more than 18 in. before grouting, and the other tier not more than 8 in. high prior to grouting. The grout is poured from the center of the course previously laid, to the center of the course being grouted. To insure grout flowing into all interstices between the bricks and to fully encase the reinforcing steel, it should be puddled immediately with a wood puddle stick about 1 in. × 2 in. In early practice, grout was puddled with the point of a trowel. This left cut planes in the grout. The fresh grout normally did not flow into the trowel penetrations, leaving voids in the grout. Thus puddling grout with a trowel is not recommended.

Grout must be well mixed before using to avoid segregation of its ingredients. It must be carefully poured, preferably with buckets equipped with spouts, to avoid having the grout contact the bed joint surfaces of exposed brickwork, and also the face of the masonry. When grout contacts the face of finished masonry it must be thoroughly cleaned off just after the mortar has set. If done sooner, the washing and scrubbing with brushes may injure the mortar joints. If done too late, the grout may have become too hard to be removed by scrubbing or washing and thus may require later sandblasting to remove it.

It is usually preferable to pour grout from the inside face of the wall when possible. Grout pouring may be done in one or more passes. By grouting from the center of one course to the center of a course just laid, it prevents movement of the top course due to hydrostatic pressure of the grout and puddling pressure.

Where grouting is suspended for an hour or more, the grouting should be carried up to within about $1\frac{1}{2}$ in. below the course or courses being grouted. This allows the grout surface to be cleaned and flushed out before proceeding with the work.

Fine grout for RGBM consists of one part portland cement and 3 parts sand which is measured in a damp loose condition.

In grout spaces 2 in. or more in width the grout may contain 2 parts pea gravel, making a mix of 1 part cement, 3 parts sand, 2 parts pea gravel and is called course grout. The addition of pea gravel not only produces a good concrete core but it also replaces water in a given volume.

When pea gravel is used it should not be pre-mixed with the sand, but rather, it should be placed in the mechanical batch mixer directly.

Where the grout space in walls, pilasters, columns and beams is wide enough to permit full brick to be inserted with at least $3/4''$ grout coverage on each side, the interior brick are "floated" into the grout for a depth of not less than $1/2$ their height.

Low Lift and High Lift Grouting

Although the terms low lift and high lift grouting have been deleted from the Uniform Building Code in recent years, they are still commonly used when referring to grouting methods.

In general, the term low lift grouting is used to refer to procedures where the grout pour height is 5 feet or less (See Table 24-G). Under these conditions UBC Section 2404(f) does not require cleanouts. However in high lift grouting, where the masonry is to be poured to heights greater than 5 feet, cleanouts are required.

UBC Table No. 24-G—Grouting Limitations

Grout Type	Grout Pour Maximum Height (Feet)[1]	Minimum Dimensions of the Total Clear Areas within Grout Spaces and Cells[2,3]	
		Multiwythe Masonry	Hollow-Unit Masonry
Fine	1	3/4	1 1/2 × 2
Fine	5	1 1/2	1 1/2 × 2
Fine	8	1 1/2	1 1/2 × 3
Fine	12	1 1/2	1 3/4 × 3
Fine	24	2	3 × 3
Coarse	1	1 1/2	1 1/2 × 3
Coarse	5	2	2 1/2 × 3
Coarse	8	2	3 × 3
Coarse	12	2 1/2	3 × 3
Coarse	24	3	3 × 4

1. See also Section 2404f.
2. The actual grout space or grout cell dimensions must be larger than the sum of the following items: (1) The required minimum dimensions of total clear areas in Table No. 24-G; (2) The width of any mortar projections within the space; and (3) The horizontal projections of the diameters of the horizontal reinforcing bars within a cross section of the grout space or cell.
3. The minimum dimensions of the total clear areas shall be made up of one or more open areas, with at least one area being 3/4 inch or greater in width.

Sequence of Grouting Masonry Walls

The sequence of grouting masonry walls depends on the type of masonry wall and height of pour or lift.

Two Wythe Brick Walls
Low Lift Grouting Procedure
12 Inches or less

When the two wythe brick walls are constructed and grouted in pours 12 inches high or less, the grout can be consolidated by puddling with a puddling stick, such as a 1" × 2" piece of wood. It is ideal for small projects in which the bricklayer grouts as he lays up the masonry units.

The following procedure for low lift grouting is an expanded version of the requirements outlined in the 1982 Edition of the UBC.

STEP 1. Concrete foundation ready to receive the mortar and first course of brick. The concrete surfaces must be clean, roughened and damp.

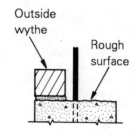

Lay one masonry course on one side with no mortar in the grout space. This wythe is called the outside wythe as it is furthest away from the mason.

STEP 2. Grout to half-height of unit and consolidate by puddling to insure grout bonding to the concrete foundation and to eliminate any voids in the grout.

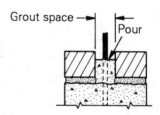

STEP 3. Lay up additional masonry units on the outside wythe up to 18 inches above the inside wythe. Projects for the California Office of State Architect limit the height of the outside wythe to 12 inches.

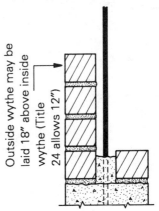

STEP 4. Grout each course as units are laid, or in pours not to exceed 6 times the width of the grout space or a maximum of 8 inches. Hold grout one inch below the top of units.

It is also permitted to lay up more than one unit, up to 12 inches high, and then pour grout and puddle. Caution and care must be taken not to displace units in grouting and puddling as the mortar is very fresh and soft.

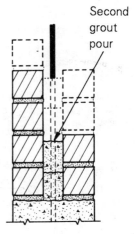

Reinforced Grouted BRICK Masonry

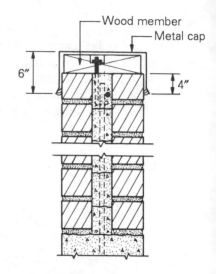

STEP 5. At completion of wall, grout flush to top of units. Puddle all grout after pouring.

More Than 12 Inches and Up To 5 Feet High

Two wythe wall may be built without cleanouts, up to 5 feet high, and grouted in one pour.

It is necessary, however, to tie the wythes together with wire ties or joint reinforcing to prevent the wythes from bulging or blowing out. In addition, the grout must be consolidated by mechanical vibration.

STEP 1. All units shall be laid with shoved heads and full bed mortar joints.

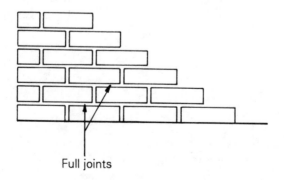

Masonry Institute of America

STEP 2. The grout space shall be not less than ³/₄ inch in width or wide enough to accommodate the vertical and horizontal reinforcing steel with proper clearance.

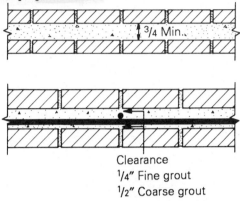

Clearance
¹/₄" Fine grout
¹/₂" Coarse grout

STEP 3. The two wythes shall be bonded together with rectangular wall ties of not less then 9 gauge wire that are 4 inches long and 2 inches less in width than the thickness of the wall. If a single zee wire tie is used, it shall not be less than ³/₁₆ inch in diameter.

Joint reinforcing may be used to tie the wythes together provided the reinforcing is spaced vertically no more than 16 inches.

There should be one rectangular tie, one zee tie or one cross-wire of joint reinforcing for approximately two square feet of wall.

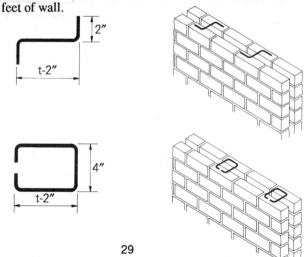

Reinforced Grouted BRICK Masonry

STEP 4. Prior to grouting, horizontal and vertical reinforcing steel, bolts and other embedded items must be in place.

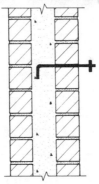

STEP 5. Give sufficient time for mortar joints to set, approximately 12 to 18 hours. This will allow joints to gain strength, and be able to withstand the pressure of the fluid grout, which may have a hydraulic head pressure up to 5 feet.

STEP 6. Grout the wall and distribute the grout evenly for the full height of the pour. Consolidate as grout is poured by means of a mechanical vibrator. Reconsolidate the grout after the excess water is absorbed by the unit, which would be approximately 3 to 5 minutes.

STEP 7. Continue to lay up masonry units, tying the wythes together and grout in each pour and consolidate by mechanical vibrator.

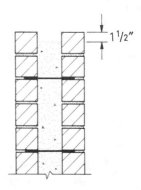

STEP 8. Horizontal construction joints must be formed by stopping all wythes at the same height. The grout is then poured to 1½ inches below the top of the masonry to form a key for the next grout pour. Where bond beams occur, stop the grout a minimum of ½ inch above the horizontal reinforcing steel. This is to insure that the horizontal reinforcing steel is covered and a shear key is provided.

STEP 9. At completion of the wall grout flush to top of units and consolidate.

High Lift Grouting Procedure

Building a wall to its full height, and then grouting, is an economical method of constructing reinforced masonry walls. It allows the mason to continually lay masonry units without waiting for the walls to be grouted. High lift grouting procedures are used when walls and grout pours exceed 5 feet.

For high lift grouting construction, cleanouts must be provided and grout must be consolidated by mechanical vibration.

STEP 1. All units shall be laid with shoved heads and full bed mortar joints.

STEP 2. The width of the grout space for walls 8 feet high shall be at least 1½ inches for fine grout and 2 inches when coarse grout is used, as a minimum, but not less than necessary to accommodate the vertical and horizontal reinforcing steel with proper clearance. For wall higher than 8 feet and up to 24 feet high, the grout space shall be wider. See UBC Table 24 G.

STEP 3. The two wythes must be bonded together with rectangular wall ties of at least 9 gauge wire that are 4 inches long, and 2 inches less in width than the thickness of the wall. If a single zee wire tie is used, it shall not be less than $3/16$ inch in diameter.

Joint reinforcing may be used to tie the wythes together, and should be spaced apart vertically, not more than 16 inches.

There should be one rectangular tie, one zee tie, or one cross-wire of joint reinforcing for approximately two square feet of wall.

Reinforced Grouted BRICK Masonry

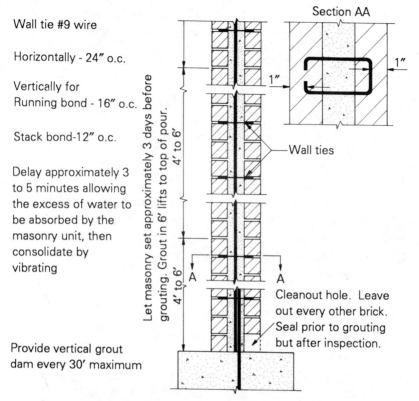

Figure 13. High lift grouting procedure for two wythe brick walls. (Cleanout holes are required)

STEP 4. Cleanouts shall be provided for each pour by leaving out brick units at the bottom of the section being poured.

STEP 5. During the work, mortar droppings and any other foreign matter shall be removed from the grout space. The cleanouts shall be sealed after inspection and before grouting.

STEP 6. Masonry walls should cure at least three days in warm weather or five days in cold weather to gain strength prior to pouring grout. This delay, which is significantly longer for high lift grouting than low lift grouting, is due to placing grout as high as 24 ft. As a result, a tremendous lateral pressure is exerted on the wythes. The use of ties will keep them from bulging or blowing out.

STEP 7.   Vertical grout barriers or dams must be built of masonry units across the grout space for the entire height of the wall to control the horizontal flow of the grout. Grout barriers must be spaced a maximum of 30 ft. apart.

STEP 8.   Grout must be a fluid mix suitable for pumping and must be mixed thoroughly. Grout can be placed by pumping or by an approved alternate method. It should be placed before any initial set occurs, and in no case more that $1\frac{1}{2}$ hours after adding the water.

STEP 9.   Grouting must be done in a continuous operation in lifts not exceeding 6 ft. It must be consolidated by mechanical vibration after excess moisture has been absorbed, approximately 3 to 10 minutes. Grouting the full height of any section of wall between control barriers should be completed in one day with no interruptions between lifts greater than one hour.

REINFORCEMENT

Before being placed, reinforcing must be free from loose rust, mill scale, and other coatings that would destroy or reduce bond.

All reinforcing must be accurately cut to length, and bent when required, without injury to the material. All kinks or bends in the bars caused by handling incident to delivery must also be straightened out without injury to the material before placement. Splices must be made only at such points, and in such manner that the structural strength of the member will not be reduced. Lapped splices must provide sufficient lap to transfer the working stress of the bars by bond, and shear with a minimum lap of 48 bar diameters for Grade 60 steel. Welded or mechanical connections must develop the design strength of the bar.

The minimum clear distance between parallel bars, except in columns, must be equal to but not less than the nominal diameter of the bars, except where lapped and tied.

All vertical reinforcing steel must be accurately placed and properly braced by mechanical devices to maintain correct positions shown on the plans. Local building codes and Title 24 C.C.R. do not require horizontal steel to be tied to vertical steel. Generally, a bricklayer can do a much better job by having the horizontal steel placed into the fresh grout than by trying to work around horizontal steel previously tied in position. Horizontal steel which is tied to vertical steel is apt to have voids under the bottom sides, but where it is placed into the grout a full solid embedment is assured.

Principal reinforcing bar spacing is limited to 4 ft. maximum each way by local codes and the Uniform Building Code; Title 24 C.C.R. limit is 2 ft. each way.

BRICKLAYING

Where feasible, brickwork should be started at a least conspicuous corner or wall, and the masonry contractor should request an early inspection of the work by the architect or engineer.

Horizontal surfaces which are to receive brickwork must be clean and damp to insure good bond between mortar and also grout and concrete. All laitance must be removed. Roughness in itself is no guarantee of providing good bond. See Title 24 C.C.R. requirements.

For maximum bond, the first course on each side of a wall should be laid on the concrete, taking care that the first mortar bed joint does not extend into the grout space, then these two courses are grouted to a point about 1½" below the top of the courses.

In laying up leads or corners, care must be taken that these leads are not too high: not over 4 feet. The most particular point to watch is the grouting of the leads. Each course in a corner lead must be grouted as laid, using a brick or other means to form a dam for the grout.

Where head joints are less than ⅝" wide, they must be solidly filled with mortar at the time of laying.

In spreading the bed mortar, extreme care must be taken to prevent the mortar from falling into the grout spaces. This can be accomplished by keeping the grout side of the mortar joint back about ½" from the inside edge of the brick so that when bricks are laid the mortar will spread to the edges of the brick in the grout space without squeezing out.

Another method, more generally used, is to spread the mortar, and then draw the trowel over the mortar in an upward and outward direction, away from the grout space forming a bevelled mortar bed. When the bricks are laid on such a bevelled bed, the mortar will likewise be squeezed only to the grout side edges without squeezing out. Deep furrowing of bed joints should not be done, however, very light furrowing or riffling the bed joint with the point of the trowel has been found to effect solid bed joints.

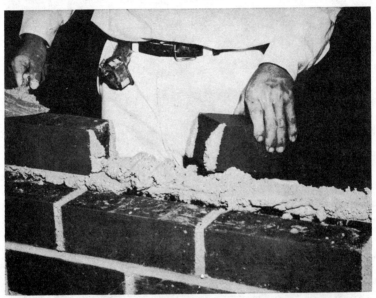

Figure 14. Insufficient head joint. Mortar is too stiff. Too much mortar in bed joint. Too much overhang in grout space. **Not recommended.**

Where the mortar squeezes out into the grout space $5/8"$ or less, either in the bed or head joint, it is best to leave it alone, as the grout will bond around it when poured and puddled. When mortar does drop into the grout space, as is bound to happen occasionally, it is best to leave it until the grouting is done, at which time the puddling of the grout will solidly mix the mortar droppings into the grout. Of course, this only applies where the grouting follows the bricklaying within a few minutes or so. Title 24 C.C.R. requires grout spaces to be kept clean of mortar droppings.

Reinforced Grouted BRICK Masonry

Figure 15. Deep furrowed bed joint. Note lengthwise void and no mortar in center of brick, and unfilled head joint. **Not recommended.**

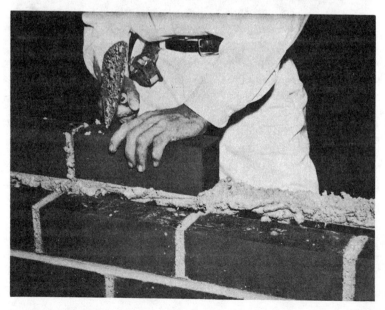

Figure 16. Furrowed bed joint. Too much overhang in grout space. **Not recommended.**

Masonry Institute of America

Figure 17. Furrowed bed joint. Too much overhang in grout space. **Not recommended.**

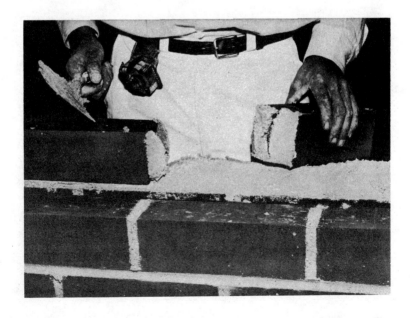

Figure 18. Full head joint. Bevelled bed joint. **Recommended.**

Reinforced Grouted BRICK Masonry

Figure 19. Full head joint. Bevelled bed joint with brick lifted. Note full coverage of mortar. **Recommended.**

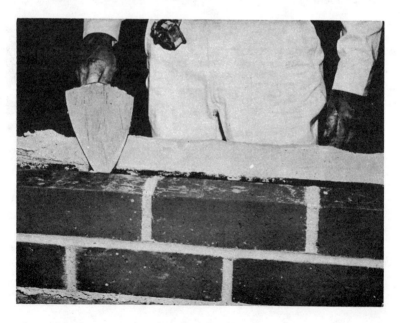

Figure 20. Bevelled bed joint. **Recommended.**

Masonry Institute of America

Figure 21. Pouring grout. Note ideal squeeze-out of mortar in back wall. See pages 26-33 for construction procedure.

Figure 22. Puddling grout with puddle stick using swishing rather than tamping procedure. **Recommended.**

Reinforced Grouted BRICK Masonry

Figure 23. Tooled joints, clean.

Figure 24. Raked and tooled joints.

Figure 25. Flush-cut joints, sack-rubbed.

Since grouting should be left down about 1 inch, below work which is to be stopped for more than an hour, there should be no mortar droppings on the top of the grout. However, before continuing work previously stopped, both the brick and the grout joint must be thoroughly cleaned and their top surfaces dampened with water.

Most bricks are required by local codes and Title 24 C.C.R. to be shoved into place. This shoving may be only a fraction of an inch, but the shoving action permits the mortar in the bed joint to be rubbed into the pores of the underside of the brick being laid and at the same time tighten up the head joints.

When necessary to stop off a longitudinal run of masonry, and at corner leads, it should be done only by racking back a half brick length, approximately, in each course and stopping the grout at the rack. Toothing must not be done without written approval of the architect or engineer.

Reinforced Grouted BRICK Masonry

Figure 26. Racking in foreground. Toothing at right background. Toothing permitted only when and where approved by architect or engineer.

Shear walls or sections are vital to a structure. Where such sections are less than about six feet in length, especially at corners, they should be built in level courses with racking or corner leads not more than 12" in height, using care to solidly grout at these corners.

An inspector is to make certain that all line pin holes and other interstices in the mortar joints are immediately filled before the masons leave an area of work.

Jointing must be done while the mortar is still thumb-print fresh plastic, so that the mortar will be squeezed tightly against the bricks before it has begun to set or stiffen. Carpet, sacks or rags **must never**

be rubbed over the face of finished exposed brickwork. Brickwork to be painted may be sackrubbed.

Lintels and beams must be complete and grouted their full height, or depth and length, in one continuous operation to avoid horizontal cold joints.

BRICK VENEER

Anchorage to Wood Frame Construction. Brick veneer over wood frame construction, with sheathing, is laid at least one inch (1″) clear from the backing. Veneer is attached to the supporting wall with corrosion-resistant metal clips of not less than 22 gauge, which are nailed into studs and plates. These clips must extend into the mortar bed joints and engage a continuous horizontal strand of wire reinforcement. This wire reinforcement is required to be not less than 9 gauge, and should be laid in the center line of the veneer bed joints. Metal clips may not support more than two (2) square feet of wall area, and may not be more than 24 inches apart horizontally.

The space between the veneer and the backing is required to be solidly filled with slush-grout as each course is laid. Dove-tail anchors may also be used.

Brick veneer over wood frame construction **without** sheathing is required to be laid at least one inch (1″) clear of galvanized metal mesh backing such as K-LATH, or STEEL TEX. To avoid moisture penetration, two layers of waterproof paper are applied directly to the studs with the metal mesh. The space between the veneer and the backing must be filled with grout as each course is laid.

Anchorage to Concrete Construction. Brick veneer is required to be laid at least one inch (1″) clear from the concrete, and is commonly attached by means of dove-tail anchorage. Dove-tail type slots may be placed not more than 24 inches apart horizontally, according to most codes. The dove-tail anchors are placed in the veneer mortar joints and should not be spaced more than 12 inches apart vertically. These anchors engage a continuous strand of horizontal reinforcement wire of not less than 9 gauge or equal, which is laid in the center line of the veneer bed joints. The space between the concrete and the backing should be filled as each course is laid.

Reinforced Grouted BRICK Masonry

VENEER

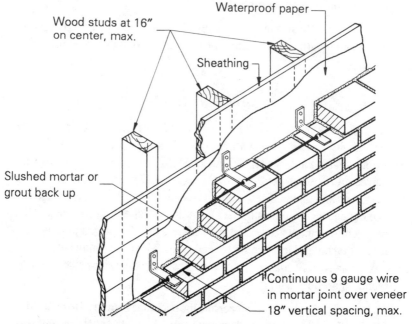

Figure 27. Anchored brick veneer to wood stud wall, Seismic Zones 3 & 4

ANCHORAGE OF VENEER

1. The veneer anchors shall be 22 gauge by 1 inch, galvanized steel brackets with 9 gauge galvanized wire.

2. Anchors shall be spaced 16 in. o.c. horizontally and not more than 18 in. vertically under the Uniform Building Code.

3. Joint wire shall be lapped at least two inches (2") on each end.

4. Veneer anchors shall be attached to studs with 8d galvanized nails.

5. Veneer backing. Slushed portland cement, mortar or grout, 1-inch thickness.

Masonry Institute of America

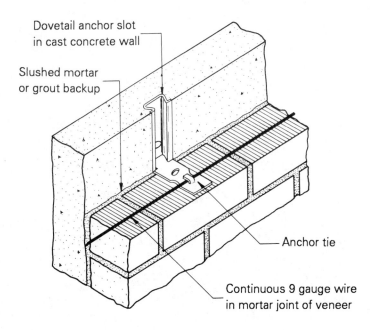

Figure 28. Anchored brick veneer to cast concrete wall.

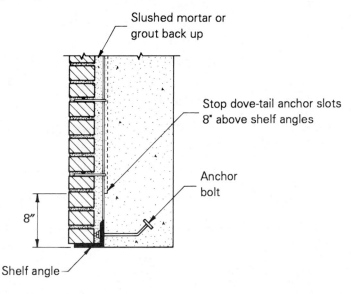

Figure 29. Shelf angle detail for brick veneer on concrete wall.

Reinforced Grouted BRICK Masonry

STEELTEX METHOD

No wood sheathing is required behind double thickness waterproofed paper backing.

A 2" by 2", 16 ga. by 16 ga. galvanized wire mesh is anchored with 2" long furring nails at 4" o.c. There is a $1\frac{1}{8}''$ penetration into each stud. At the top and bottom plates 8d common nails at 8" o.c. can be used.

For brick veneer over steel stud framing, wire Aqua K-Lath or Steeltex can be used to metal studs or channels with standard tie wires.

AQUA K-LATH METHOD

No wood sheathing is required behind the self-furring galvanized welded wire fabric lath, which is 16 gauge, $1\frac{1}{2}'' \times 2''$ mesh, with waterproof paper laminated to the back side.

The K-lath is secured to wood studs with galvanized barbed nails at 6" on center at furring crimp, $1\frac{1}{8}''$ penetration. At the top and bottom plates 8d common nails at 8" o.c. should be used.

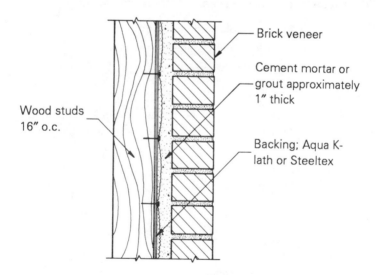

Figure 30. Anchored veneer, no wood sheathing back up required, using Steeltex or Aqua K-lath backing.

Masonry Institute of America

Figure 31. Wood sheathing, wood studs, using dove-tail slots and dove-tail anchors which engage wire reinforcement, being slush grouted. See details pages 44, 45 and 46.

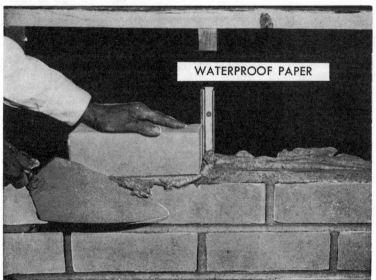

Figure 32. Racking in foreground. Toothing at right background. Toothing permitted only when and where approved by architect or engineer, except brick being laid in full mortar bed which embeds wire and anchors.

Reinforced Grouted BRICK Masonry

Figure 33. "Steeltex" or "K-lath" on wood studs, being slush grouted.

Figure 34. "K-lath" on wood studs, being slush grouted.

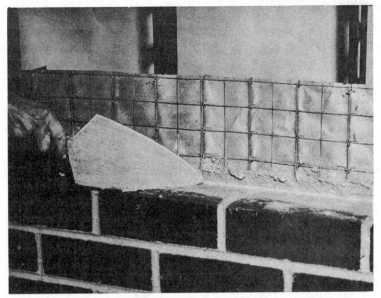

Figure 35. "Steeltex" tied to steel studs, being slush grouted.

BRACING, SHORING AND SCAFFOLDING

All scaffolding must be of safe design and not overloaded by materials and be in accordance with national or local safety rules.

Scaffolding can not be used to brace walls and must be maintained independent of the walls.

PROTECTION OF WORK

Faces of masonry must be protected from being smeared or splattered with mortar, grout or splashings from scaffolds. However, should this occur, the faces should be cleaned with clear water and fiber brushes immediately after the initial set of the mortar in the joints—about 2 hours. All sills, ledges and offsets of brick or other materials must be protected from mortar droppings during construction. It is wise to protect the tops of all unfinished masonry from rain. No construction supports should be attached to masonry unless specifically permitted by the architect or engineer.

CURING OF FINISHED MASONRY

There is more than enough water in RGBM construction to completely hydrate the cement. It is not recommended to wet down brick masonry during or after construction, except in extreme hot, dry weather where there is evidence that the bricks are absorbing water too fast from the mortar which can be seen readily as wet edges around the brick adjacent to the mortar joints, (see Wetting of Brick). In such cases, the masonry walls should have their surfaces dampened only by a light fog spray, without water running down the face of the work, after the mortar has set.

CORING OF WALLS

Where walls are required to be cored, the cores should be taken at least six (6) feet from a corner, and preferably in a full wall, rather than under a window. Normally, two (2) cores on a non-state project are sufficient. See Core Tests, Title 24 C.C.R.

TEST SPECIMENS OF MORTAR AND GROUT

ASTM specification C 270, "Specifications for Mortar for Unit Masonry", contains requirements for masonry mortar based on properties of mortar, as well as on proportions of the ingredients of mortar by volume, the former being presumed to meet the requirements of the latter and vice versa. This ASTM Specification requires that the test specimens be prepared in a laboratory. In this specification, mortar is not required to meet both the property and the proportion specifications.

The following procedures are those which are outlined in Title 24 C.C.R. and the Uniform Building Code, 1991 edition:

> 1. Mortar used in masonry construction must be classified in accordance with the materials and proportions set forth in Table 24-A, Aggregate for mortar must conform to ASTM C 144 per UBC Section 2402(b)1A.
>
> 2. Admixtures and mortar colors may not be added to the mortar unless approved by the building official. Provisions for such additives must be provided in the contract specifications. After

admixtures are added to the mortar, it must conform to the requirements of the property specifications.

Only pure mineral oxide, carbon black or synthetic colors may be used.

3. The strength of mortar using cementitious materials set forth in Table No. 24-A should meet the minimum compressive strength shown in UBC Standard No. 24-20. The building official may require field tests to verify compliance with this section. Such tests must be made in accordance with UBC Standard No. 24-22.

Note, however, that UBC Standard No. 24-20 is based on ASTM C 270, which in Paragraph 7.1 of the 1989 version states: "Specification C 270 is not a specification to determine mortar strengths through field testing."

4. Mortar mixed to a flow suitable for use in laying masonry units is generally not required to meet the laboratory flow.

Reinforced Grouted BRICK Masonry

UBC TABLE 24-A MORTAR PROPORTIONS FOR UNIT MASONRY

Mortar	Type	Proportion by Volume (cementitious materials)									Aggregate Measured in a Damp Loose Condition
		Portland Cement or Blended Cement[1]	Masonry Cement[2]			Mortar Cement[3]			Hydrated Lime or Lime Putty[1]	
			M	S	N	M	S	N		
Cement–lime	M	1	—	—	—	—	—	—	1/4	Not less than 2 1/4 and not more than 3 times the sum of the separate volumes of cementitious materials
	S	1	—	—	—	—	—	—	over 1/4 to 1/2	
	N	1	—	—	—	—	—	—	over 1/2 to 1 1/4	
	O	1	—	—	—	—	—	—	over 1 1/4 to 2 1/2	
Mortar cement	M	1	—	—	—	—	—	—	—	
	M	—	—	—	—	1	—	—	—	
	S	1/2	—	—	—	—	—	—	—	
	S	—	—	—	—	—	1	—	—	
	N	—	—	—	—	—	—	1	—	
Masonry cement	M	1	—	—	—	—	—	—	—	
	M	—	1	—	—	—	—	—	—	
	S	1/2	—	—	—	—	—	—	—	
	S	—	—	1	—	—	—	—	—	
	N	—	—	—	1	—	—	—	—	
	O	—	—	—	1	—	—	—	—	

1. When plastic cement is used in lieu of portland cement, hydrated, lime or putty may be added, but not in excess of one tenth of the volume of cement.
2. Masonry cement conforming to the requirements of UBC Standard No. 24-16.
3. Mortar cement conforming to the requirements of UBC Standard No. 24-19.

Masonry Institute of America

Special Seismic Requirements for Zones 2, 3 and 4. U.B.C. Section 2407 (h) 3,4

3. Special provision for Seismic Zone No. 2. Masonry structures in Seismic Zone No. 2 shall comply with the following special provisions.

A. Materials. The following materials shall not be used as part of the structural frame: Type O mortar, masonry cement, plastic cement, nonload-bearing masonry units and glass block.

B. Wall reinforcement. Vertical reinforcement of at least 0.20 square inch in cross-sectional area shall be provided continuously from support to support at each corner, at each side of each opening, at the ends of walls and at a maximum spacing of 4 feet apart, horizontally throughout the wall.

Horizontal reinforcement not less than 0.2 square inch in cross-sectional area shall be provided: (1) at the bottom and top of wall openings and shall extend not less than 24 inches nor less than 40 bar diameters past the opening, (2) continuously at structurally connected roof and floor levels and at the top of walls, (3) at the bottom of the wall or in the top of the foundations when dowelled to the wall, (4) at maximum spacing of 10 feet unless uniformly distributed joint reinforcement is provided. Reinforcement at the top and bottom of openings when continuous in the wall may be used in determining the maximum spacing specified in item No. (1) above.

C. Stack bond. Where stack bond is used, the minimum horizontal reinforcement ratio shall be .0007 bt. This ratio shall be satisfied by uniformly distributed joint reinforcement or by horizontal reinforcement spaced not over 4 feet and fully embedded in grout or mortar.

4. Special provisions for Seismic Zones Nos. 3 and 4. All masonry structures built in Seismic Zones Nos. 3 and 4 shall be designed and constructed in accordance with requirements for Seismic Zone No. 2 and with the following additional requirements and limitations.

EXCEPTION: One-and two-story structures of Group R, Division 3 and Group M Occupancies is Seismic Zone No. 3 with h'/t not greater than 27 and using running bond construction may be constructed in accordance with the requirements of Seismic Zone No. 2.

Reinforced Grouted BRICK Masonry

When half allowable stresses are used in the f'_m from Table No 24-C shall be limited to a miximum of 1500 psi for concrete masonry and 2600 psi for clay masonry unless f'_m is verified by tests in accordance with Section 2405(d)3A and 3D or 3F. A letter of certification is not required.

When half allowable stresses are used for design, the f'_m shall be limited to 1500 psi for concrete masonry and 2600 psi for clay masonry for Section 2405(d)1C and 2E unless f'_m is verified during construction by the testing requirements of Section 2405(d)1B. A letter of certification is not required.

Reinforced hollow-unit stacked bond construction which is part of the seismic resisting system shall use open-end units so that all head joints are made solid, shall use bond-beam units to facilitate the flow of grout and shall be grouted solid.

FIELD TEST SPECIMENS OF MORTAR AND GROUT

While the preceding pertains to unit masonry, similar conditions occur when using grout in structural reinforced brick masonry.

Figure 36. Spreading mortar on masonry unit.

Masonry Institute of America

Figure 37. Mortar specimen and cardboard mold.

Field Test of Mortar, refer to UBC Standards No. 24-22

Compressive Test Specimens of Mortar. Spread the mortar on masonry units being used between $1/2''$ and $5/8''$ thick, and allow to stand for one (1) minute, then remove mortar and place in a $2'' \times 4''$ cylinder in 2 layers, compressing the mortar into the cylinder using a flat end stick or using fingers. Lightly tap mold on opposite sides, level off molds, immediately cover, and keep them damp until taken to the laboratory. After 48 hours, have the laboratory remove specimen from the molds and place the mortar specimens in the fog room until tested in the damp condition.

Field Test of Grout, refer to UBC Standards No. 24-28

Slump of Grout. Use a truncated metal cone $4''$ dia. to $8''$ dia., $12''$ high, and after dampening, place on a flat moist, non-absorbent base. Thoroughly mix or agitate grout before placing in cone to obtain a full representative mix. Fill cone with grout, lightly tap cone on opposite

Reinforced Grouted BRICK Masonry

sides, level off, carefully lift cone and measure slump. At time of placing, grout should have a slump of 9" plus or minus 1".

Compressive Test Specimens of Grout. On a flat non-absorbent base, form a space approximately 3" × 3" × 6" high, using bricks having the same moisture condition as those being laid. Line the space with a permeable paper or porous separator so that water may pass through the liner into the brick. Thoroughly mix or agitate grout to obtain a full representative mix. Place grout into molds in 2 layers and puddle each layer with a 1" × 2" puddling stick to eliminate air bubbles. Level off and immediately cover molds and keep them damp until taken to the laboratory. After 48 hours, have the laboratory carefully remove bricks, and place the grout specimens in the fog room until tested in the damp condition.

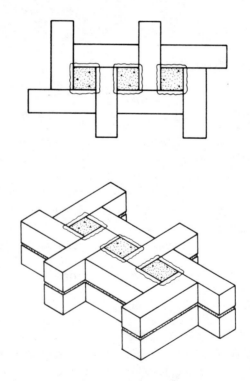

Figure 38. Brick molds lined with paper towels, for test specimens of grout, approximately 3" × 3" × 6" high.

Field Test Values for Compressive Strength

Following are minimum mortar and grout compressive strength requirements according to Title 24 C.C.R. and UBC Standard 24-22.

	28 day Compressive Strength (psi)
MORTAR— 1 Part portland cement, $\frac{1}{2}$ part lime, and $4\frac{1}{2}$ parts dry loose sand	1500
GROUT— 1 part portland cement, 3 parts dry loose sand with from 1 to 2 parts pea gravel	2000
CORES Cut from Walls (Title 24 C.C.R.)	1500

EFFLORESCENCE

Efflorescence. This stubborn problem has caused confusion and trouble for masonry since its first appearance thousands of years ago on ancient masonry walls. Efflorescence is the white, powdery scum that often appears on masonry walls after construction. It is undesirable and unwanted, but a persistent problem.

A great deal of time, money and effort have been spent trying to solve the difficulties efflorescence generates. Many test programs have been developed and numerous attempts have been made to eliminate the efflorescence problem. Unfortunately, nothing has proven 100% effective against this very stubborn problem. However, even though no surefire cure has been discovered, a great deal has been learned about how efflorescence works and how to prevent it, and if preventive measures were inadequate, how to remove the efflorescence if it does appear.

What is Efflorescence

Efflorescence is a fine, white, powdery deposit of water-soluble salts left on the surface of masonry as the water evaporates. The efflorescent salt deposits tend to appear at the worst times, usually about a month after the building is constructed, and sometimes as long as a year after completion.

Reinforced Grouted BRICK Masonry

Figure 39. Typical white efflorescent salts on brick masonry.

Required conditions

Efflorescence is not a simple subject. Three conditions must exist before efflorescence will occur.

First: There must be water-soluble salts present somewhere in the wall.

Second: There must be sufficient moisture in the wall to render the salts into a soluble solution.

Third: There must be a path for the soluble salts to migrate through to the surface where the moisture can evaporate, thus depositing the salts which then crystalize and cause efflorescence.

All three conditions must exist. If any one of these conditions is not present, then efflorescence cannot occur.

Even though the efflorescence problem is complex, it is not difficult to prevent. Although no economically feasible way exists to totally eliminate any one of these three conditions, it is quite simple to reduce all three and make it nearly impossible for efflorescence to occur.

CLEANING

The traditional method of cleaning has been sandblasting, which of course, works; unfortunately it removes just about everything else. The abrasive action of the sand erodes the surface of the brick and the tooled mortar joints along with any deposited salts. This increases the porous qualities of the masonry and the water absorptive nature of the wall. Sandblasting will also damage the integrity of the dense tooled mortar joints. A well-tooled and compacted mortar joint readily sheds moisture and provides minimum voids for penetration. After sandblasting, the mortar is more porous, has void for infiltration, and may even reveal cracks in the mortar. Additionally, the appearance of the masonry wall will be changed since the texture of the brick has been made slightly coarser.

Sandblasting should be use with caution and afterwards the masonry should be sealed with a waterproofing material.

A conventional chemical cleaner that has been used for removing efflorescence is muriatic acid in a mild solution, usually one part muriatic acid, hydrochloric acid, HC1, to 12 parts water. Several mild individual applications are better than one overpowering dose. Again, care must be taken to thoroughly presoak the wall with clean water and to thoroughly flush the wall of all remaining acids with clean water.

Cleaning efflorescence from masonry walls does not cure the problem, it only removes the symptoms. After cleaning, the efflorescence will reappear unless the natural efflorescent chain is broken. Due to the added water used when pre-soaking and post-flushing the walls when using chemical or acid cleaners, the efflorescence will sometimes reappear, often stronger than before.

Efflorescence is a controllable condition that should not be a problem in modern masonry. Breaking the chain of conditions necessary for efflorescence can be done with attention to details, the correct materials and quality construction.

PAINTING AND WATERPROOFING

When grouted brick masonry and brick veneer walls are properly constructed, with tooled mortar joints, they are invariably waterproof, do not leak, and require no water-repellent material to exposed surfaces.

It requires many weeks of warm dry weather for a grouted brick wall to become absolutely dry throughout its interior. However, most free water comes to the surface as vapor within a few months, depending on weather conditions of temperature and humidity.

Caution

No paint or sealant should be applied to a grouted brick masonry wall until a lapse of several months after completion, or following the rainy season, or a hard rain at other times. The walls should be virtually dry. When paint or a water-repellent is to be applied to exterior brickwork, the work should be done only by qualified applicators at least ten days before landscaping is installed, and before sprinklers are turned on. Sprinklers should always be faced away from brickwork to avoid intermittent soaking of the masonry.

Before the application of any paint or water-repellent material to brick surfaces, it is recommended that a qualified representative of the manufacturer of the proposed coating material inspect the work, and verify the type of material to be used, also checking the moisture, temperature and surface conditions of the masonry. It is best that brick surfaces have a temperature of not less than 40 degrees F. at time of painting or waterproofing. Only qualified and experienced applicators should be used to apply the material or materials.

Surface inspection reports can be obtained from paint or sealant manufacturers or applicators and in this event their recommendations should be followed to the letter.

Weather conditions play an important part in painting and waterproofing. Brick construction which has been thoroughly saturated during construction, and completed during the rainy season offers a puzzling painting problem. After a few days of warm weather the surfaces appear relatively dry; however, interior areas of the wall are invariably still wet.

Many specifications are written in the belief that it is mandatory to paint or waterproof the exteriors of brick walls prior to plastering or painting the interior surfaces. If this is done, the presence of efflorescent salts may not be too apparent, but as time goes on a disruption of the paint film by these efflorescent salts might occur. If weather conditions permit, the painting or waterproofing of the exterior surfaces should be delayed as long as possible.

Too much emphasis cannot be made on the fact that successful and lasting applications of paint or water-repellent coatings on brick masonry can be obtained only if applied after the masonry has dried out.

Before a paint or water-repellent is applied to a wall, the surface must be cleaned of all dirt, oil, grease and efflorescence, if any, and be thoroughly dry. If cement-water paint is to be used and the wall has been previously painted, such paint must be entirely removed. Soaps or detergents must not be used in cleaning since they will prevent adhesion of the paint unless they are completely removed from the wall surface. All cracks over $1/32$ inch wide must be filled or sealed separately before colorless water-repellent materials or paints are applied.

COMPOSITION OF COATINGS
The following are "breather" type coatings:

CEMENT-WATER paints are water reducible and usually contain about 80% white portland cement, and about 20% silica sand passing the 40 mesh sieve. Proprietary brands may, in addition, contain small percentages of titanium dioxide and a water-repellent, usually a stearate. Cement-water paints do not contain an organic binder. Resin-emulsion, oil-base and synthetic-rubber paints are referred to as organic paints.

ACRYLIC-RESIN emulsion paints are water reducible pastes of pigments ground in an acrylic vehicle which contains no oil and which have been treated with an emulsifying agent to render them miscible with water.

POLY-VINYL ACETATE emulsion paints are water reducible formulations of selected paint pigments ground in a vehicle of high polymer poly-vinyl acetate emulsion as a binder.

SILICONE-BASE coatings are those having a silicone resin non-volatile (solid) content of not less than 5% by weight. Mineral spirits or other aromatic solvents are greatly preferred to water as a vehicle for this type of coating.

ELASTOMERIC coatings are also used as both a block filler and paint finish in one application. They are usually much more water resistant than conventional paints or textured coatings and have the advantage of being able to bridge hairline cracks and still maintain integrity when the cracks open up due to shrinkage in the wall. In order to waterproof a cracked or porous wall, however, these coatings must be applied with a sufficiently thick application in order to fill or cover all the pores. Patching over cracks with an elastomeric sealant is usually required before the coating can be applied. Finish appearance retains somewhat more of the original masonry textured coatings but less than the clear or stains.

The following are "sealer" (vapor barrier) type coatings:

SYNTHETIC-RUBBER paints are of two types: (1) The rubber-solution type in which the synthetic rubber is added to a vehicle of treated drying oils, aromatic hydrocarbons, and coal tar thinners; and (2) the rubber-emulsion type, may be breather or sealer-type, in which the synthetic-rubber resin is treated with an emulsifying agent so that the paste paints are reducible with water.

OIL-BASE paints are ready mixed, containing opaque pigments suspended in a vehicle of drying oils and thinner.

BITUMINOUS COMPOUNDS are asphalt or coal tar which are usually applied hot. Certain asphalt compounds and asphalt emulsions are applied cold. Surfaces must be THOROUGHLY DRY at time of application.

Parapet and Fire Walls

The most important parts of a structure to be waterproofed, aside from the roof, are the tops and inner exposed surfaces of parapet walls or fire walls above the roof. After the masonry has thoroughly dried, these surfaces should be sealed with a coat of heavy asphalt type MASTIC compound applied from the roofing or flashing up the walls and over the tops of such walls. A positive seal may be obtained by using a fibered asphalt, or trowel-type compound of the cut-back type similar to "Insul-Mastic No. 4010" or "Sika Seal", Trowel Grade, rather than an emulsion type, applied in accordance with the manufacturer's specifications by a qualified waterproofing applicator. These are Gilsonite-Asphalt compounds, naturally black, but which may be colored by using ceramic granules sprayed or rolled into the wet mastic to produce almost any desired color. Lighter colors up to white may be obtained by using an acrylic resin type paint over the black compound after the compound has dried for about a week.

Capping of Walls and Sills

Where cement is used to cap tops of walls or sills, it should be applied not sooner than $1/2$ hour after mixing, and the capping should be made as thin as practicable, about $1/4''$. The surfaces to be capped must be dampened before mortar is applied and, after application, the capping must be kept moist for at least 24 hours.

When header bricks are used for capping, the die-skin or edge surfaces of the brick, not the wire-cut surfaces, should be exposed to the weather and the top and end joints be tightly tooled.

A good recommended procedure is to use a metal coping over the top of a parapet wall with the roofing material carried up under the metal coping.

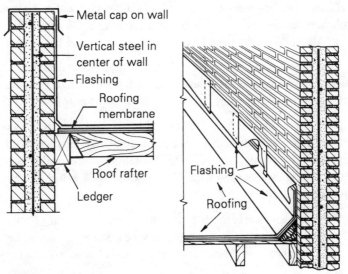

Figure 40. Parapet walls, cap and flashing detail.

Reinforced Grouted BRICK Masonry

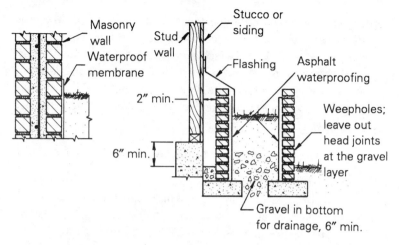

Figure 41. Planter box details. Planter boxes attached to wood framed structures are required to have a clearance of not less than 2 inches from the face of the wall, whether of stucco, wood siding, wood sheathing or wood studs.

Planters

All types of masonry planters whether wood or stucco, including concrete, attached to wood frame structures, are required to have a clearance of at least 2 in. from the face of a wall. Alkali salts present in the water, earth, and fertilizers are a cause of efflorescence on this type of brickwork. All mortar joints must be tool jointed on tops and both sides.

Planters may have small weep holes through vertical joints of the lowest exposed course. All four inside surfaces of planters should be back plastered with mortar, and then water-proofed as follows:

Before any earth, or gravel is placed in planter boxes, it is recommended that a coat of heavy mastic compound, as mentioned under Parapet and Fire Walls, be applied to all interior surfaces. Allow mastic waterproofing in planters to stand for at least one week before gravel and earth fill is placed.

It should be remembered that earth in planters "builds up," or swells, over a period of time, thus allowance must be made for the future top of the soil. Finished soil should be at least 6 in. below an adjacent floor level.

PLASTERING

Masonry surfaces on which suction must be reduced, should be wetted down 48 hours and again 24 hours before plastering operation start.

Masonry joints may be struck off flush. Surfaces must be free from paint, oil, dust, dirt or other material that might prevent satisfactory bond.

The surface may be given a cement-grout dash coat, composed of 1 part cement to $1\frac{1}{2}$ parts fine sand mixed with water to make a workable consistency, or a plaster bonding agent of a non-oxiding, non-crystallizing, resinous water emulsion (non-bituminous products).

Surfaces of masonry walls must be thoroughly and evenly dampened, but not saturated, immediately preceding the application of the scratch coat. The surfaces must appear slightly damp but there must be no free water on them. The plaster mortar must be forced tightly against the surfaces.

Cement-grout dash-bond coat may take the place of a scratch coat. For more data on plastering using different materials and finishes, see:

Contracting Plasterers Association of Southern California
(213) 690-5721

Western Lath, Plaster, Dry Wall Industries Assoc.
1700 Green Briar Lane #260
Brea, CA 92621

Plastering Information Bureau
21243 Ventura Blvd. #115
Woodland Hills, CA 91364

SCHOOLS, HOSPITALS AND PUBLIC BUILDINGS, TITLE 24
CALIFORNIA CONSTRUCTION REQUIREMENTS

The construction of public schools in California is required to comply with regulations of the Field Act as published in Title 24 of California Construction Requirements (C.C.R.). Many of its requirements are more rigid than those of most building codes. Following is a list of some of them affecting RGBM.

Reinforced Grouted BRICK Masonry

Cement, only Type I or Type II portland cement (ASTM C 150) may be used.

Sand, ASTM C 144 with not less than 2% passing the No. 100 sieve, may be washed or unwashed for grout, and unwashed for mortar.

Mortar, 1 part portland cement, $\frac{1}{4}$ to $\frac{1}{2}$ part lime, and 4 parts sand based on dry loose volume. Allowance for bulking is to be made for damp sand. No admixtures are permitted unless approved by the Office of the State Architect. Accurate box measurement of quantities, no shovel measurements.

Grout, 1 part portland cement, 3 parts sand with not less than 1 part or more than 2 parts pea gravel based on dry loose volume. Grout space, $2\frac{1}{2}''$ minimum. Approved admixtures may be added. Accurate measurement of quantities, no shovel measurements.

Mortar strength. Minimum compressive strength of $2'' \times 4''$ cylinders, 1500 psi at 28 days. Test mortar is taken from masonry soon after spreading (not from mortar boards).

Grout strength. Minimum compressive strength of $3'' \times 3'' \times 6''$ specimens made in brick molds, 2000 psi at 28 days, tested in vertical position. Grout to be taken from grout tub.

Absorption of brick. At time of laying, bricks shall have a residual total absorption between 5 and 15 percent.

Steel spacing. Maximum 2 feet each way, minimum splice 48 diameters. Grout cover over steel, minimum $\frac{1}{2}$ inch.

Bricklaying. Start of work; (see drawing, page 26 and 28) step 1, concrete surfaces must have aggregate exposed, clean and damp; step 2, lay one course on one side, no mortar in grout space; step 3, lay one course on other side, no mortar in grout space; step 4, grout to half-height of brick and puddle to insure bond to concrete; step 5, lay one side up not over 12" high; step 6, lay courses on other side, puddling each course as laid. Keep mortar out of grout spaces. Keep brickwork damp for 3 days after laying.

Full head joints, brick shoved into place. Bed joints flat or bevelled up and out from grout spaces.

Column ties minimum $3/8''$ diameter, not less than $1\tfrac{1}{2}''$ from face, not $8''$ apart.

Bolt coverage not less than one inch (1″) of grout. Increase of 50% in allowable shear values for vertical bolts when loads are applied at top of and parallel to walls. All bolts must be accurately set with templates including bolts in tops of walls. Vertical bolts may not be forced in place into previously poured grout. Grout space may not be left open for length of bolts but must be grouted to top level of wall or pilaster before leaving.

Masonry Core Tests
Excerpt from Title 24 .C.C.R. Section 2405(c)4.C.

C. Masonry Core Tests. Not less than two cores having a diameter of approximately two-thirds of the wall thickness shall be taken from each project. At least one core shall be taken from each building for each 5,000 square feet of floor area or fraction thereof. The architect or structural engineer in responsible charge of the project or his representative (inspector) shall select the areas for sampling. One-half of the number of cores taken shall be tested in shear. The shear loading shall test both joints between the grout core and the outside wythes and webs of masonry. Core samples shall not be soaked before testing. Materials and workmanship shall be such that for all masonry when tested in compression, cores shall show an ultimate strength at least equal to the f'_m assumed in design but not less than 1,500 pounds per square inch.

For more detailed requirements refer to Title 24 C.C.R. and the Uniform Building Code Sections 2402 through 2405.

Figure 42. Compression test on core cut from wall.

Reinforced Grouted BRICK Masonry

Figure 43. Testing shear strength, grout to brick.

EARTHQUAKE CONSIDERATIONS

No part of the surface of the earth is free from earthquakes. No one can predict earthquakes. Since earthquakes are vibrations transmitted as waves in the materials of the earth, they may originate from a variety of causes, including landslides, explosive volcanic activity, movements of molten rock at depth, natural or artificial explosions and abrupt movements of masses of rock along faults in the outer part (crust) of the earth. In California, the most famous fault is the San Andreas.

Although earthquakes are a hazard, it is interesting to note that for the past century less than 1000 people have been killed in all the United States and Canada (including those killed by the fire following the San Francisco quake in 1906) as a result of earthquakes; or 1 death in 5 million.

The most destructive force is caused by horizontal earth motion. When the ground underneath a structure is moved suddenly to one side, the building will tend to remain in its original position because of its inertia. Thus all parts of the building must be so connected to the frame that the building will vibrate as a unit, the connections being strong enough to overcome the inertia of the separate parts. When adjacent buildings have different periods or amplitudes, which is generally the case, they should be separated sufficiently to keep them from hammering one another during an earthquake.

Details of all connections must be thoroughly checked as per plans, particularly the tightening of bolts. Eccentric connections should be given special attention.

For further information on earthquakes, see publications by:

Structural Engineers Association
of Southern California
2550 Beverly Blvd.
Los Angeles, CA 90057; (213) 385-4424

Structural Engineers Association
of Northern California
217 Second Street
San Francisco, CA 94105; (415) 974-5147

Structural Engineers Association
of San Diego
P.O. Box 26500, Suite 203
San Diego, CA 92126; (619) 223-9955

Structural Engineers Association
of Central California
P.O. Box 19440
Sacramento, CA 95816-1959; (916) 965-1536

Reinforced Grouted BRICK Masonry

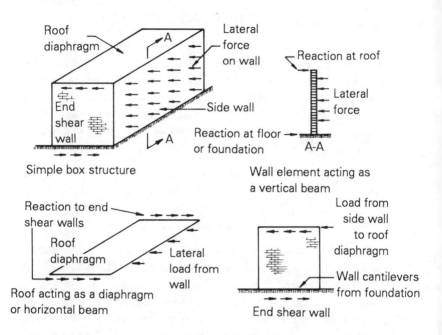

Figure 44. Lateral forces on a box structure.

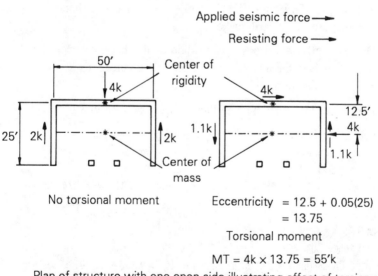

Plan of structure with one open side illustrating effect of torsion

Figure 45. Seismic forces on buildings.

Masonry Institute of America

FIREPLACE AND CHIMNEY NOTES

Verify that a chimney which is not built into a masonry wall does not support any load other than its own weight.

Check footings for thickness and minimum projection beyond chimney walls.

Check thickness of walls, with or without flue lining, and thickness of separation where more than one flue is contained in the same chimney.

Check mortar, grout, bricklaying and specifications.

Check fireplace opening and hearth.

Check fill-in of solid masonry behind fire-back up to about 6" below top of throat to form a depressed smoke chamber.

Check flue and throat areas.

Check minimum horizontal of ½" between top of throat and inner face of flue.

Check corbelling, size and spacing of reinforcement, bond beams, anchorages, separation from combustibles, also damper and spark arrester if required.

Check height of chimney above roof.

Check F.H.A. and V.A. requirements if required.

For complete details, obtain Residential Masonry Fireplace and Chimney publication from the Masonry Institute of America, 2550 Beverly Blvd., Los Angeles, California 90057-1085.

A masonry fireplace and chimney are a beneficial energy conservation device for any home. Locating the fireplace inside, away from exterior walls, provides a thermal mass that will store and release heat energy.

During the winter when the fireplace is in use, the masonry will store much heat energy and release it slowly through the night when the surrounding air cools down.

The masonry fireplace acts as a heat sink that creates a thermal lag and reduces the extreme highs and lows of internal home temperature.

Reinforced Grouted BRICK Masonry

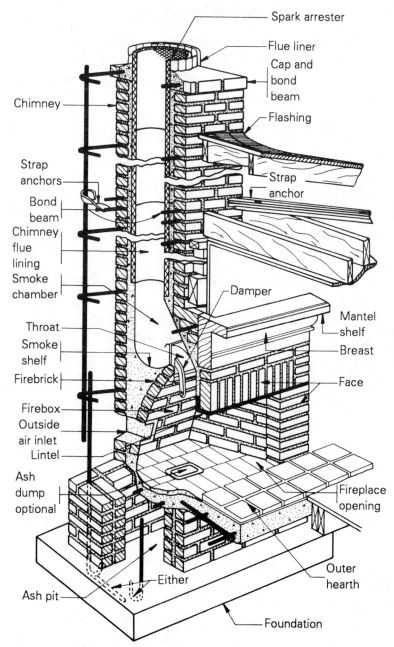

Figure 46. Parts of a fireplace and chimney.

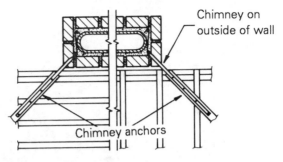

Where plates are cut, anchor to chimney by steel straps hooked into chimney and attached to plates by two $3/8'' \times 3''$ lag screws; or two $1/2''$ bolts

Figure 47. Anchorage of chimney

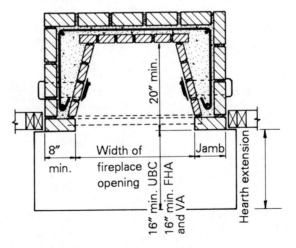

Figure 48. Plan of fire box at top of hearth.

Reinforced Grouted BRICK Masonry

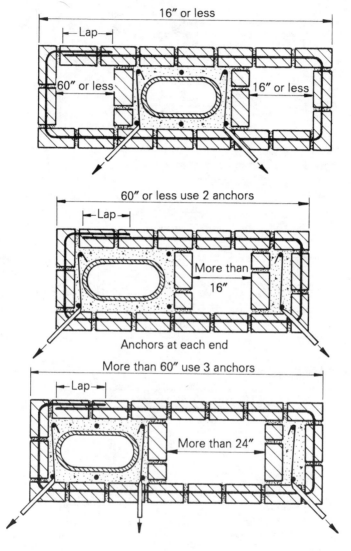

Figure 49. Anchorage of wide chimney

Masonry Institute of America

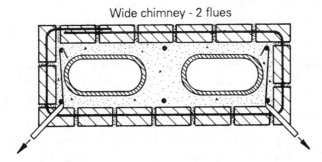

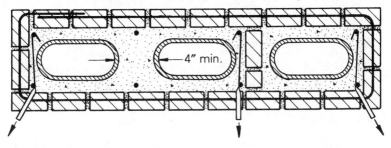

Figure 50. Anchorage of chimney with multiple flues.

TIMELY TIPS FOR INSPECTION

A vital part of any construction project is good inspection. The inspector's job is important. Knowledge and good judgement are essential to obtaining the results required by the approved plans and specifications. The materials furnished on the job represent the manufacturers' efforts to supply products meeting job specifications. It is the inspector's responsibility to see that these products are properly used to produce a quality job.

Prior to starting masonry construction, the inspector must verify that necessary material testing has been performed as required. Some tests may have to be conducted well in advance of job site delivery such as for high strength block. All materials must meet specified requirements.

Reinforced Grouted BRICK Masonry

The inspector should keep a daily log from his first day on the project. The status of the job from the beginning should be noted.

The daily log should record weather, temperature, and job conditions. The inspector should record all materials, test specimens and job progress and note what work was accomplished and where it was done. This includes laying of masonry units and grout pours that are completed.

It is also suggested that the inspector note how many masons are on the job each day and the delivery of materials. Any special conditions, problems or adverse events that may take place should be noted.

If there are job conferences, a list of who attended, what was accomplished, and the decisions made should also be noted.

Complete and thorough job records are very important, and the inspector is invaluable in maintaining these.

As with all competent and skilled professionals and craftsmen, construction inspectors must have their tools and materials to properly carry out their inspection duties and responsibilities. The following is a minimum suggested list an inspector should have.

1. A current set of plans and specifications, including all changed orders.
2. Applicable building codes and standards to which the project was designed and the requirements of the jurisdiction it is under.
3. A list of architects, engineers, contractors and subcontractors; names, addresses, telephone numbers and responsible persons.
4. A notebook or log to keep daily notes as described above.
5. Necessary forms for filing reports with required agencies.
6. Pens, pencils and erasers.
7. Folding rule or retractable tape and long steel tape.
8. String to check straightness.
9. Keel — yellow, blue and black.
10. Permanent felt tip markers for labeling specimens.

11. Hand level and plumb bob.

12. Small trowel and smooth rod for making and rodding mortar and grout samples.

13. Sample molds obtained from testing laboratory.

14. Absorbent paper towels and masking tape to take grout specimens.

There can be more items needed, depending on the project and the scope of duties required of the inspector.

Brick. Check brick for compliance and specifications.

Mortar. Study the UBC section on mortar carefully, also the section on Title 24 C.C.R. requirement if applicable. Check if laboratory tests are required for cement. Check volume measurements and bulking of sand, (see Section on Sand). Check shovel measurements by using a cement sack for one cubic foot and check occasionally during the job. Check order and time of mixing. It is very important to see that mortar is maintained highly plastic or 'soft' on the mortar board. Refer to bottom of page 19 on "Mortar" for proper retempering.

Grout. Grout should be thoroughly mixed before placement in the wall.

Steel. Check if laboratory tests are required. Check position and length of dowels. Steel must be checked constantly for position and clearance from brick. When coring is required, note the location of steel so it can be avoided when cutting cores.

Bricklaying and Grouting. Don't interfere with workmen or their work. Keep in the clear when inspecting. Give instructions only to foremen. Don't insist on wetting brick which are rain-soaked. check uniformity of width of vertical and horizontal mortar joints. Report careless bricklayers to the foreman.

See that head joints are actually full with mortar or grout. See that grout solidly fills the grout space. Where excessive longitudinal shrinkage cracks occur between grout pours, check dampness of brick and amount of water in grout, and use of lime and pea gravel, and make adjustments.

Reinforced Grouted BRICK Masonry

See that shear walls less than 6' wide, particularly at corners, are built level the full length with racking or corner leads not more than 12" high, and see that such leads are solidly grouted. See that jointing with jointer tool is done while mortar is still plastic enough to be pressed against edges of brick. Mortar used to form caps on walls or sills should be used not sooner than $1/2$ hour after original mixing to prevent shrinkage cracks. Tops of masonry must be damp while such caps are placed, and be kept covered and damp for at least 24 hours.

See that corners of finished masonry are protected against damage, especially where materials or equipment are handled through or around them. See that line pin holes, and other holes between masonry and other materials are properly filled at completion.

It is far better for an inspector to be a cooperating helper in seeing that the plans and specifications are properly executed, than an antagonistic troublemaker. Bricklaying is considered an art. Artists are temperamental and so are bricklayers. However, with proper conduct, an inspector can make his own work and that of the bricklayers a cooperative, congenial experience, for the benefit of all concerned.

LAP SPLICES FOR REINFORCING STEEL

It is not reasonable to build a reinforced masonry wall using a single continuous length of reinforcing steel. Instead, the steel is placed using bars that have been cut to manageable lengths. For these shorter lengths of steel to function as continuous reinforcement, they must be connected in some fashion.

The usual method it to lap the bars a specified length. The Uniform Building Code requires that reinforcing steel bars in tension have a lapped length of 40 bar diameters for Grade 40 steel and 48 bar diameters for Grade 60 steel. When the steel is in compression the required lap length is 30 bar diameters for Grade 40 steel and 36 bar diameters for Grade 60 steel.

Splices may be made only at certain locations and in such manner that the structural strength of the member will not be reduced.

Masonry Institute of America

UBC SEC. 2409(e)6

The amount of lap of lapped splices shall be sufficient to transfer the allowable stress of the reinforcement as in Section 2409(e)3. In no case shall the length of the lapped splice be less than 30 bar diameters for compression and for Grade 40 diameters for tension.

Welded or mechanical connections shall develop 125 percent of the specified yield strength of the bar in tension.

TABLE A
Length of Lap (inches)[1]
Grade 40 Steel, F_s = 20,000 psi

Bar Size		Laps for Compression Bars			Laps for Tension Bars		
No.	Dia., d_b (inches)	30 Dia.[2] Min.	30x1.3[3] = 39 Dia.	30 x 1.5[4] = 45 Dia.	40 Dia.[2] Min.	40 x 1.3[3] = 52 Dia.	40 x 1.5[4] = 60 Dia.
3	0.375	12[5]	15	18	15	20	23
4	0.500	15	20	23	20	26	30
5	0.625	19	24	29	25	33	38
6	0.750	23	29	35	30	39	45
7	0.875	26	34	39	35	46	53
8	1.000	30	39	45	40	52	60
9	1.128	34	44	51	45	59	68
10	1.270	38	50	57	51	66	77
11	1.410	42	55	63	56	73	84

1. Based on UBC Sec. 2409(e)6 and 2409(e)3.G.
2. Minimum development length, ld, and lap splice.
3. Use when bars are separated by 3 inches or less.
4. Use where f_s>0.80F_s per UBC Sec. 2409(e)3.G.
5. Min.lap length = 12" recommended.

Reinforced Grouted BRICK Masonry

TABLE B
Length of Lap (inches)[1]
Grade 60 Steel, F_s = 24,000 psi

Bar Size		Laps for Compression Bars			Laps for Tension Bars		
No.	Dia., d_b (inches)	36 Dia.[2] Min.	36x1.3[3] = 47 Dia.	36 x 1.5[4] = 54 Dia.	48 Dia.[2] Min.	48 x 1.3[3] = 62 Dia.	48 x 1.5[4] = 72 Dia.
3	0.375	14	18	21	18	23	27
4	0.500	18	24	27	24	31	36
5	0.625	23	29	35	30	39	45
6	0.750	27	35	41	36	47	54
7	0.875	32	41	48	42	54	63
8	1.000	36	47	54	48	62	72
9	1.128	41	53	62	54	70	81
10	1.270	46	60	69	61	79	92
11	1.410	51	66	77	68	87	102

1. Based on UBC Sec. 2409(e)6 and 2409(e)3.G.
2. Minimum development length, ld, and lap splice.
3. Use when bars are separated by 3 inches or less.
4. Use where f_s>0.80F_s per UBC Sec. 2409(e)3.G.
5. Min.lap length = 12" recommended.

The Uniform Building Code Sec. 2409(e)6 requires that when adjacent bars (2 or more) are spaced 3 inches or closer apart, the required lap length must be increased 30%.

Therefore for compression, 36 bar diameters (db) laps (Grade 60 steel) would be required to be increased to 47 db and 48 db for tension would be increased to 62 db, if 2 or more bars are lapped at the same location. See Table above.

If the lapped splices are staggered, the lap lengths need not be increased.

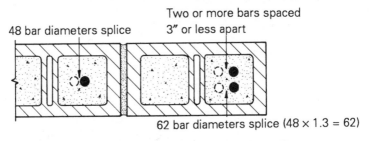

Figure 51. Lap splice of steel in cell.

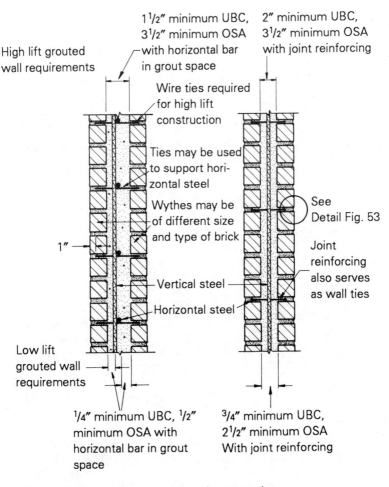

Figure 52. Wall clearances for grouted construction.

Reinforced Grouted BRICK Masonry

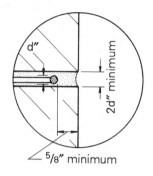

Figure 53. Detail of edge cover and clearance of joint reinforcing.

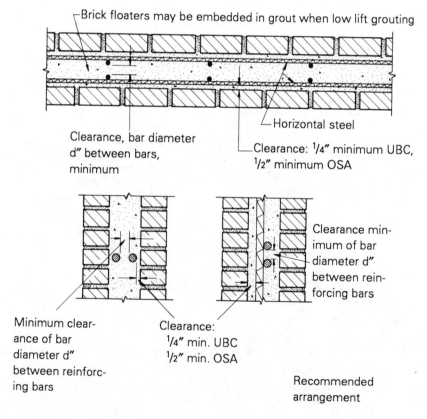

Figure 54. Clearances of steel in a beam or wall.

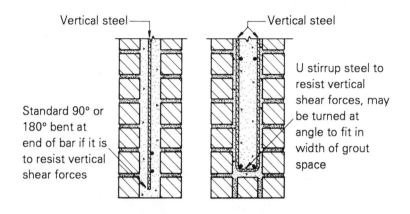

Figure 55. Vertical shear reinforcing steel arrangement for beams.

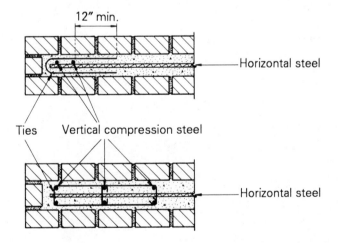

Figure 56. Ties for compression or jamb steel at the end of walls or piers.

Reinforced Grouted BRICK Masonry

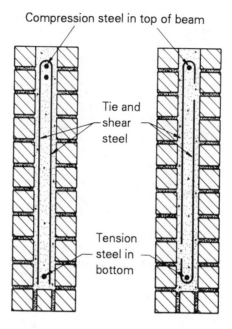

Figure 57. Ties for compression steel in beams.

FLUSH WALL COLUMNS

If engineering design permits, it is to the economic benefit of the owner and to the construction benefit of the contractor to build columns that are contained in the wall and are flush with the wall. Wall-contained columns permit faster construction, since there are no projections from the wall and no special units are required. The reinforcing steel must be tied in accordance with the code requirements.

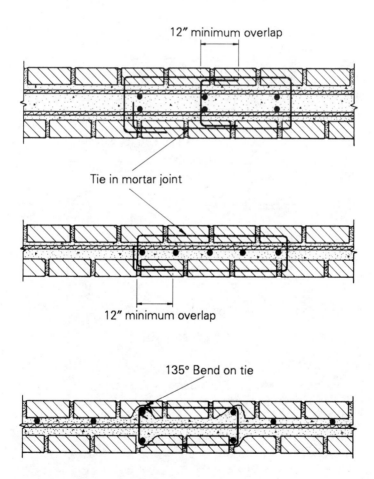

Figure 58. Flush wall brick columns with ties in mortar joint.

Reinforced Grouted BRICK Masonry

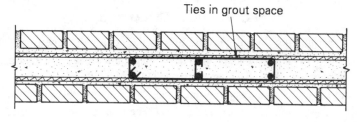

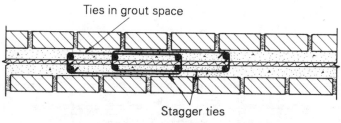

Figure 59. Flush wall brick columns with ties in grout space.

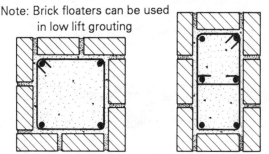

Figure 60. Brick columns showing vertical steel and ties.

COLUMN TIES

Excerpt from Section 2409(B)5 1991 UBC

5. Reinforcement for columns. A. Vertical reinforcement. The area of vertical reinforcement shall be not less than .005 Ae and not more than .004 Ae. At least four No. 3 bars shall be provided.

B. Lateral ties. All longitudinal bars for columns shall be enclosed by lateral ties. Lateral support shall be provided to the longitudinal bars by the corner of a complete tie having an included angle of not more than 135 degrees or by a hook at the end of a tie. The

Masonry Institute of America

corner bars shall have such support provided by a complete tie enclosing the longitudinal bars. Alternate longitudinal bars shall have such lateral support provided by ties and no bar shall be farther than 6 inches from such laterally supported bar.

Lateral ties and longitudinal bars shall be placed not less than $1\frac{1}{2}$ inches and not more than 5 inches from the surface of the column. Lateral ties may be against the longitudinal bars or placed in the horizontal bed joints if the requirements of Section 2407(f) are met. Spacing of ties shall be not more than 16 longitudinal bar diameters, 48 tie diameters or the least dimension of the column but not more than 18 inches.

Ties shall be at least $\frac{1}{4}$ inch in diameter for No. 7 or smaller longitudinal bars and No. 3 for larger longitudinal bars. Ties less than $\frac{3}{8}$ inch in diameter may be used for longitudinal bars larger than No. 7, provided the total cross-sectional area of such smaller ties crossing a longitudinal plane is equal to that of the larger ties at their required spacing.

Table C   Tie Spacing—16 Bar Diameters

Compression Steel Bar No.	Maximum Tie Spacing
3	6"
4	8"
5	10"
6	12"
7	14"
8	16"
9	18"
10	18"
11	18"

Note: Maximum tie spacing: 16 bar diameters or 18" or least dimension of column.

Reinforced Grouted BRICK Masonry

Table D   Tie Spacing—48 Tie Diameters

Tie Steel Bar No.	Maximum Tie Spacing
1/4″	12″
3	18″
4	18″
5	18″

Note: #2 (1/4″) ties at 8″ spacing equivalent to #3 (3/8″) tie at 16″ spacing. Maximum tie spacing is 48″ tie diameters or 18″ or least dimension of column.

> C. **Anchor bolt ties.** Additional ties shall be provided around anchor bolts which are set in the top of the column. Such ties shall engage at least four bolts or, alternately, at least four vertical column bars or a combination of bolts and bars totaling four in number. Such ties shall be located within the top 5 inches of the column and shall provide a total of 0.4 square inch or more in cross-sectional area. The uppermost tie shall be within 2 inches of the top of the column.

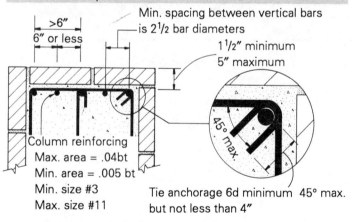

Figure 61. Reinforcing tie details for Seismic Zones 3 and 4.

Lateral ties shall be placed not less than 1 1/2 inches and not more than 5 inches from the surface of the column, and may be against the vertical bars or placed in the horizontal bed joints where permitted by Section 2407(g).

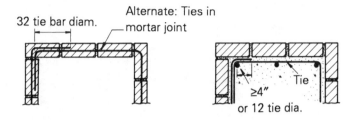

Figure 62. Tie details for Seismic Zones No. 0, 1 and 2.

Where the ties are placed in the horizontal bed joints, when permitted by Section 2407(g), the hook shall consist of a 90-degree bend having a radius of not less than four ties diameters plus an extension of 32 tie diameters.

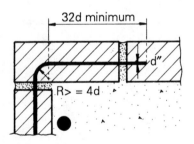

Figure 63. Tie anchorage in bed joint Seismic Zones 3 and 4.

In Seismic Zones 0, 1 and 2, no intermediate ties need be provided for bars located in between the corner bars. There is no limiting distance either to the location of intermediate bars between the corner compression bars.

Reinforced Grouted BRICK Masonry

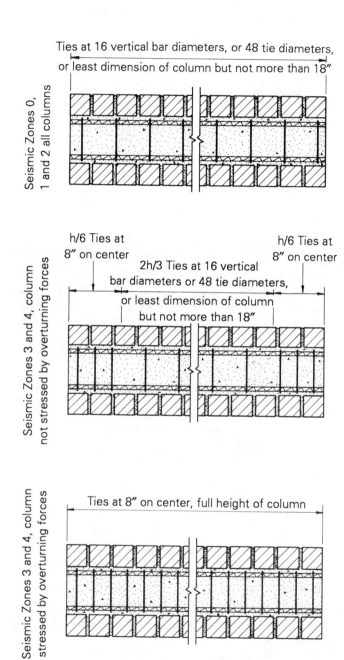

Figure 64. Tie spacing in columns.

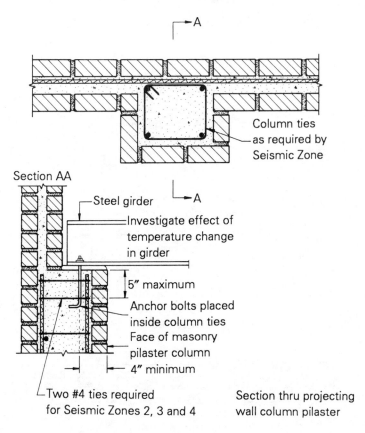

Figure 65. Projecting wall column pilaster.

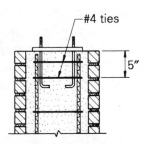

Figure 66. Ties at anchor bolts on top of columns in Seismic Zones 2, 3 and 4.

Details of Brick Construction

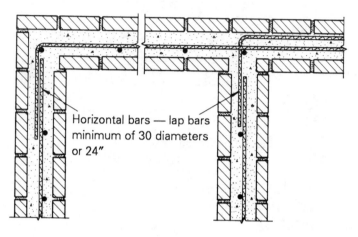

Single curtain of steel

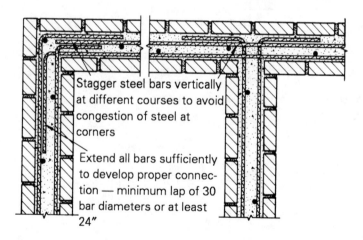

Double curtain of steel

Figure 67. Typical wall connections.

Masonry Institute of America

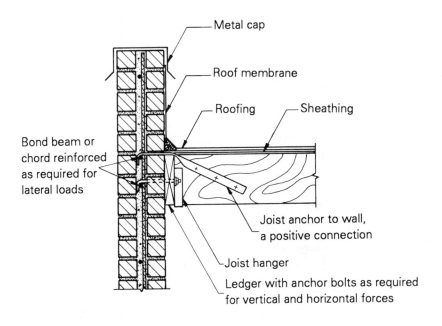

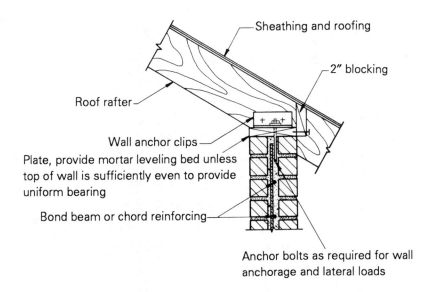

Figure 68. Connection details to wood roof.

Reinforced Grouted BRICK Masonry

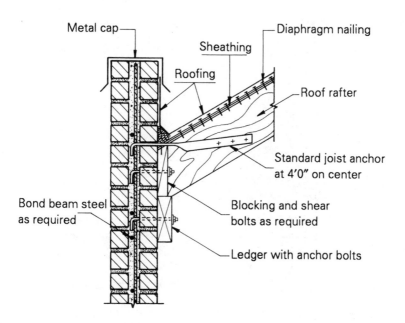

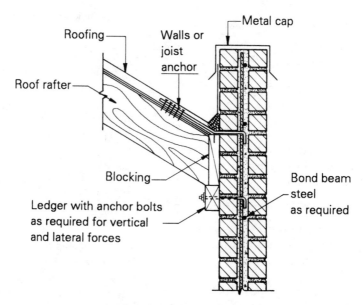

Figure 69. Connection detail to wood roof.

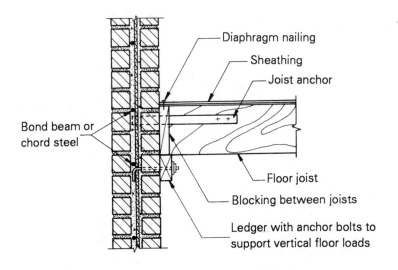

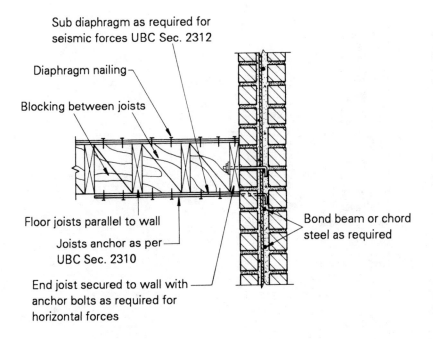

Figure 70. Connection details to wood floor.

Reinforced Grouted BRICK Masonry

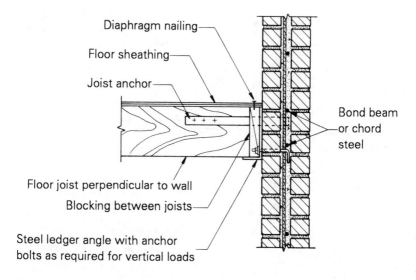

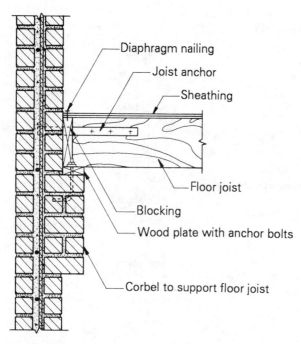

Figure 71. Connection details to wood floor.

Masonry Institute of America

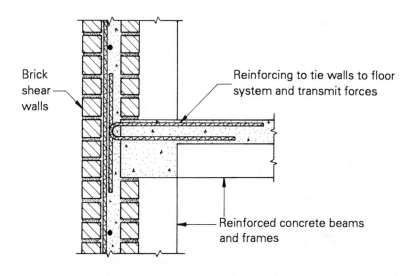

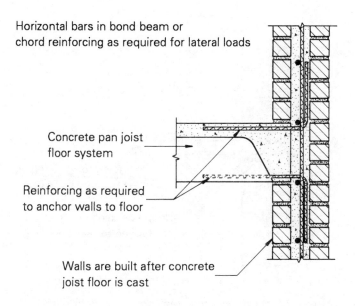

Figure 72. Masonry to concrete connection details.

Reinforced Grouted BRICK Masonry

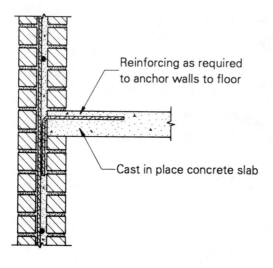

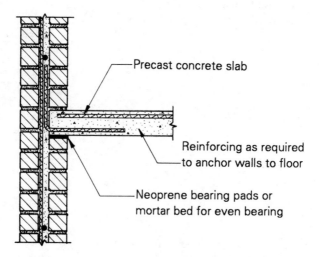

Figure 73. Masonry to concrete connection details.

Masonry Institute of America

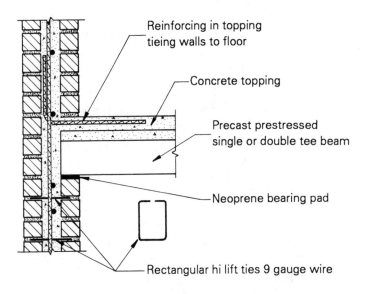

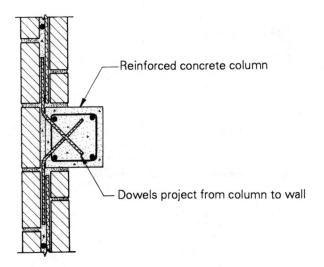

Figure 74. Masonry to concrete connection details.

Reinforced Grouted BRICK Masonry

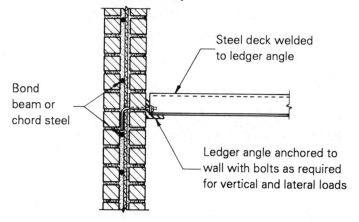

Figure 75. Masonry to steel connection detail.

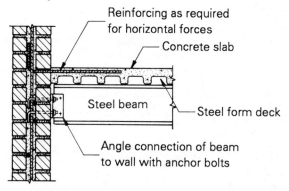

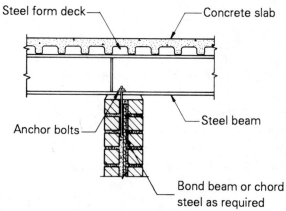

Figure 76. Steel to masonry connection details.

Masonry Institute of America

GLOSSARY OF TERMS RELATING TO BRICK MASONRY

Absorption. The amount of water, by weight, absorbed by a brick when totally immersed in water, expressed as a percentage of its dry weight. Absorption is determined by immersion in cold water for 24 hours or in boiling water for 5 hours.

Absorption Rate, Initial Rate of Absorption, or Rate of Suction. The amount of water, by weight, absorbed by a brick in one minute when held in 1/8 inch of water, expressed in grams or ounces per sq. inch of the lower surface.

Admixtures. Materials added to mortar or grout as water repellent or to retard or speed up setting.

ASTM. American Society for Testing Materials.

Backup. Basically, that part of a masonry wall which is behind the exterior facing. Backup masonry may be finished facing. Specifications should clearly indicate types and locations of backup materials.

Bat. A piece of brick, usually ½ brick or less.

Batter. Recessing or sloping a wall or buttress back in successive courses; the opposite of corbelling.

Bearing Partition or Wall. A wall which supports a vertical load in addition to its own weight.

Bed Joint. The horizontal layer or mortar on or in which a masonry unit is laid.

Bond. (Adhesion) The tensile strength between brick and mortar and grout.

Bond. (Pattern) The arrangement of units to provide strength and stability.

    English Cross (or Dutch) **Bond.** Alternate courses of stretchers and headers offset 2" by corners starting with 2", 4", 6", 4", 2", etc. Stretchers continue from 2" and 6" corners, headers continue from 4" corners. Used with standard size brick.

Reinforced Grouted BRICK Masonry

Flemish Bond. Courses of alternating stretcher, header, stretcher, header ... with headers centered over stretchers in alternate courses. Used with standard size brick.

Running Bond. Brick laid so that the vertical mortar joints are centered over the brick below. Sometimes called center bond or half bond.

Stack Bond. Vertical mortar joints in continuous vertical alignment. Also called plumb joint bond, straight stack, jack bond and jack on jack.

Third Bond and Quarter Bond. Brick laid so that the wall corners show alternate headers and stretchers, and the stretchers continue from this point and the vertical mortar joints occur over about one-third or one-fourth the length of the stretchers.

Bond Beam. The course or courses of masonry units reinforced with longitudinal bars and designed to take the longitudinal flexural and tensile forces that may be induced in a masonry wall. It usually occurs at floor and roof levels.

Breaking Joints. The arrangement of masonry units so as to prevent continuous vertical joints in adjacent courses.

Brick:

Building Brick. A solid masonry unit made from clay or shale, dried and burned or fired at about 1850 degrees F., usually without special attention to color with smooth, textured, or designed surfaces, as made in Southern California in various sizes.

Cored Brick. Bricks which have holes (any shape) through them. When the gross cross-sectional area of cores does not exceed 25% of the cross-sectional area of a brick, it is classified by ASTM as a solid unit.

Dry-Pressed Brick. Brick produced by a process of forming relatively dry clay in molds at pressures of from 550 to 1500 psi.

Face Brick. A solid masonry unit made from a scientific and controlled mixture of natural clays or shales only, and specially processed during manufacture to produce desired colors and textures, regularity in shape and uniformity, and made in modular sizes. Also called **Facing Brick.**

Fire Brick. Brick made from refractory clay which resist heat shock and high temperatures; not cored.

Paving Brick. A brick which is not cored.

Salmon Brick. A relatively soft underburned brick, so called because of its color.

Sewer Brick. A brick, low in absorption, high in compressive strength, which resists action of sewer gas and abrasion.

Soft-Mud Brick. Brick produced by a process of forming relatively wet clay in molds. When the inside of the molds are sanded, the product is called "sand-struck" or "sand-mold" brick. When molds are wetted with water to prevent sticking of the clay, the brick are called "water-struck."

Stiff-Mud Brick. Brick produced by extruding a plastic clay through a die.

Buttering. Placing mortar on brick with a trowel before laying.

Cavity Wall. Not permitted by codes in areas subject to earthquake probability. A wall built of masonry units constructed to provide a continuous center air space, usually 2 inches wide, within the wall. The wythes, or tiers, or masonry are tied together with rigid non-corrosive metal ties at specified intervals.

Cementitious Materials. Portland cement, plastic or waterproof cement, masonry cement, and lime. Low alkali portland cement is recommended to reduce efflorescence.

Ceramic Veneer. A type of architectural terra cotta, characterized by larger face dimensions and thinner sections ranging from $1\frac{1}{8}$ in. to $2\frac{1}{2}$ in. in thickness.

Adhesion type. Ceramic slabs approximately $1\frac{1}{8}$ in. in thickness (Title 24 C.C.R., 1 in. maximum), held in place by the adhesion of the mortar to the ceramic veneer and to the backing wall. No metal anchors are required.

Anchored Type. Ceramic slabs approximately 2 to $2\frac{1}{2}$ in. in thickness held in place by wire anchors and a grout space in which vertical pencil rods are placed. The slabs are anchored to the rods, which in turn, are anchored to the backing wall.

Chase. A continuous recess built into a wall to receive pipes, ducts, etc.

Closer. The last brick laid in a course. A closer may be a whole unit or one that is shorter and usually appears in the field of the wall.

C/B Ratio. See **Saturation Coefficient.**

Collar Joint. The interior longitudinal vertical joint in a brick masonry wall. In RGBM it is the grout space.

Column. A vertical isolated structural compression member whose ratio of unsupported length to least width is greater than four.

Coping. The material or units used to form a cap or finish on top of a wall, pier or pilaster to protect the masonry below from penetration of water from the above.

Corbel. A shelf or ledge formed by projecting successive courses of masonry out from the face of the wall or an off-set in vertical alignment as in an offset chimney.

Course. One of the continuous horizontal layers of masonry bonded together with mortar, forming the masonry structure.

Cross Joint. See **Head Joint.**

Curtain Wall. A non-bearing wall built between columns or piers for the enclosure of a building.

Cut Sides or Faces. The cut sides or cut faces of building brick are the two sides which have been cut by cutting wires as the clay column is extruded from the brick making machine. The distance between the cutting wires determines the thickness of the bricks. The other four sides are the die-skin sides whether they be smooth or textured. Most Norman or Roman face brick in Southern California are end-cut, meaning the bricks are extruded and cut lengthwise. Additionally, 3 of the long sides are normally wire-cut or ruffled while the bottom side remains smooth as it rides along the conveyor belt.

Dwarf Wall. A wall or partition which does not extend to the ceiling.

Efflorescence. A whitish powder, sometimes found on the surface of masonry or concrete due to deposits of water soluble salts.

Exfoliation. Scaling or flaking of the surfaces.

Facing. Any material forming an integral part of a wall, used as a finishing surface contrasted with veneer which is not integral with the wall.

Faced Wall. A wall in which the facing and backing are so bonded with masonry, or otherwise tied, as to exert common action under load.

Fat Mortar. A mortar that tends to be sticky and adheres to the trowel.

Fire Clay. A finely ground refractory clay. Varies widely in physical properties.

Fire Wall. Any wall which subdivides a building so as to resist the spread of fire, by starting at the foundation and extending continuously through all stories to, or above, the roof.

Flaking. See **Exfoliation**.

Flow or Mortar. Measured on a steel circular plate upon which mortar has been placed into a small truncated cone, the cone removed, and the plate "dropped" 25 times in 15 seconds. The "flow" is the increase in diameter of the mortar pat over the original diameter of the base of the cone. (Laboratory work). Mortar for masonry should have a flow between 135% and 145% at time of use.

Flow After Suction. First the "flow" is determined, then the mortar is placed in an apparatus whereby a 2" vacuum is pulled on the mortar through a perforated dish for one minute. Then the mortar is again measured for "flow." Flow after suction is expressed as a percentage of original flow (Laboratory work). Good mortar should have not less than 85% flow after suction at time of use.

Frog. A depression in the bed surface of a brick. Sometimes called a **panel**.

Furrowing. The practice of striking a "V"-shaped trough in a bed of mortar.

Grade SW (Severe Weather). Brick intended for use where a high degree of resistance to frost action is desired and the exposure is such that the brick may be frozen when permeated with water.

Reinforced Grouted BRICK Masonry

Grade MW (Moderate Weather). Brick intended for use where exposed to temperatures below freezing but unlikely to be permeated with water, or where a moderate and somewhat non-uniform degree of resistance to frost action is permissible.

Grade NW (No Weather). Brick intended for use as back-up or interior masonry.

Grout. A mixture of portland cement, sand and water, with or without the addition of pea gravel or lime, poured into place in masonry.

Harsh Mortar. A mortar that, due to an improper measure of materials, is difficult to spread.

Head Joint. The vertical mortar joint between ends of masonry units; sometimes called the **Cross Joint**.

Header. A Masonry unit laid flat with its greatest dimension at right angle to the face of the wall. Generally used to tie two wythes of masonry together. Not permitted in RGBM Construction.

High Lift Grout. See pages 31 to 33.

Hog (in a wall). A wall that is hogged is one in which, for any given height at both ends, there is a different number of courses, necessitating adjustment of mortar bed joints or the use of supplementary height units in the field.

Hollow Tile. Masonry units in which the coring exceed 25% of the gross cross-sectional area of the unit.

Hydrated Lime. Quicklime to which, in manufacture, only sufficient water is added to stabilize the lime's chemical affinity for water, and is used in powdered form. Water added to hydrated lime makes lime putty.

Jointer. A metal tool used to finish the surfaces of mortar joints, producing concave, flat, V, beaded, etc., joints.

Jointing. The process of finishing mortar joints with a trowel, jointer, rubber ball, etc.

Kiln Run. Building bricks from one kiln which have not been sorted or graded for size or color variation.

Lead. The section of a wall built up and racked back on successive courses at the corners or end of a wall. The line is attached to the leads and the wall then built up between them.

Lean Mortar. A mortar that is low in cementitious material.

Lime Putty. Quicklime to which sufficient water has been added to make a plastic paste ready for use in mortar.

Line-Pin Holes. Holes left by pins to hold bricklayers coursing lines.

Lintel. A horizontal structural member placed over an opening in a wall to carry the superimposed weight above.

Modular Masonry Unit. A masonry unit whose nominal dimensions are based on the 4 in. module.

Mortar. A plastic mixture of cementitious material, sand and water to bind bricks together and to prevent water penetration.

Nominal Dimension. A dimension which may vary from the actual dimension of the unit by the thickness of a mortar joint.

Panel Wall. A non-bearing wall in skeleton frame multi-story construction, built between columns or piers, and wholly supported at each story.

Parapet Wall. That part of an exterior wall entirely above the roof line.

Parging. The process of applying a coat of cement mortar to the back of the facing material or the face of the backing material; sometimes referred to as Pargeting.

Party Wall. A wall used for joint serviced by adjoining buildings.

Pea Gravel. Mineral aggregate, well graded, with not more than 5% passing the No. 8 sieve and with 100% passing the $3/8''$ sieve.

Pick and Dip. A method of laying brick by which the bricklayer simultaneously picks up a brick with one hand and, with his other hand, enough mortar on the trowel from the mortar board to lay one brick. Usually done only on special ornamental design work.

Pier. A vertical isolated structural compression member whose ratio of unsupported length to least width is four or less. A pier whose width is less than three times its thickness must be designed and constructed as required for a column.

Pilaster. A portion of a wall which projects on one or both sides of a wall and acts as a vertical beam, column or both.

Plasticizing Agent. Material which will cause mortar or grout to become plastic or workable. Lime is not only the best plasticizing material for mortar or grout but is also a cementitious material.

Pointing. Filing mortar into a joint or hole in masonry after the masonry unit is laid.

Processed Lime. Pulverized quicklime.

Quarter Bond. See **Third Bond.**

Racking. A method of building the end of a wall by stepping back each course, so that it can be built onto and against other brick courses without toothers; also used in corner leads.

Raggle. A grove or channel in a mortar joint, or in a special masonry unit (raggle block), to receive roofing, flashing or other material which is to be sealed into the masonry.

Rate of Absorption. See **Absorption Rate.**

Reinforced Grouted Brick Masonry. (RGBM). Reinforced brick masonry in which the continuous longitudinal vertical or collar joint is filled with grout and reinforcing steel. No masonry headers are used in this type of construction.

Retempering. To moisten mortar and re-mix, after original mixing, to the proper consistency for use.

Rowlock. A brick laid on its face edge, laid in the wall with its long dimension at right angle to the wall face. Frequently spelled rolok.

Saturation Coefficient. A measure of the resistance to freezing and thawing of a brick. It is the ratio of the weight of water absorbed by cold immersion in 24 hours to the weight absorbed by immersion in boiling water in 5 hours. Not required in Southern California, except in high mountainous areas subject to frost action.

Shiner. A brick laid on its edge so that its greatest dimension is horizontal and parallel to the face and its wider side exposed.

Shoved joints. Produced by placing a brick on a mortar bed and then immediately shoving it a fraction of an inch horizontally against the mortar in the head joints to effect solid, tight joints.

Slushed Joints. Collar joints filled after units are laid by "throwing" mortar in with the edge of a trowel.

Soap. A brick manufactured or cut on the job so that its width is less than standard. Usually supplied by manufacturer in $1/2$ standard width.

Soldier. A brick laid on its end so that its greatest dimension is vertical and its edge exposed.

Solid Masonry Unit. The great majority of brick in this area are actually solid; however, bricks are established in ASTM specifications as being solid when core holes or hollow spaces do not exceed 25% of the bearing area of the brick as normally laid. Tests have shown that such cored bricks are as strong as non-cored bricks made in the same way from the same clay. The steel coring pins around which the clay passes while being extruded from the brick machine, tightens or squeezes the surrounding clay, hence such cored bricks are usually of equal compressive strength to those not cored.

Spall. A small fragment removed from the face of a masonry unit by a blow or otherwise.

Spandrel Wall. That part of a panel wall above the top of a window in one story and below the sill of the window in the story above.

Story Pole. A pole marked by the bricklayer showing each course and used to measure vertical heights during the construction of the wall.

Stretcher. A masonry unit laid with its longest dimension horizontal and parallel to the face of the wall, with its edge exposed.

Stringing Mortar. The procedure of spreading enough mortar on the bed joint to lay several masonry units.

Struck Joint. Any mortar joint which has been finished smooth with a trowel.

Temper. See **Retempering.**

Tier. Each vertical 4 inches or single unit section or thickness of masonry; also called **withe** or **wythe.**

Reinforced Grouted BRICK Masonry

Tolerance. ASTM Standards for Building or Face Brick.

Tooling. Compressing and shaping the face of a mortar joint with a special tool other than a trowel.

Toothing. The temporary ending of a wall wherein the units in alternate courses project in vertical alignment. Should be only permitted in structural walls when approved by Architect or Engineer.

Tuck Pointing. The filling in with fresh mortar of cut-out or defective mortar joints in masonry.

Veneer. Exposed masonry attached to backing but not structurally bonded to the backing to sustain loads.

Vitrified. That characteristic of a clay product resulting when the temperature in the kiln is sufficient to fuse all the grains and close all the pores of the clay, making the mass impervious.

Voids. Empty spaces in sand, mortar, grout, etc.

Water Retentivity. An important property of a mortar which prevents its rapid loss of water into masonry units. (See text on **Flow after Suction and Water Retentivity**.)

Wire Cut Sides. See **Cut Sides.**

Wythe or Withe. See **Tier.**

Masonry Institute of America

GUIDE SPECIFICATIONS FOR REINFORCED GROUTED BRICK MASONRY

NOTE: These guide specifications follow Uniform Building Code requirements. Some modifications may be required by other codes.

Scope of Work

Work Included:

1. The work under this section includes all labor, materials, equipment, tools and appliances required to complete the brick masonry work as indicated on the drawings and as herein specified.

2. Pointing and cleaning of masonry.

3. Waterproofing and sandblasting of masonry where specified.

4. Removal of surplus masonry materials and waste upon completion of the masonry work.

5. Welding as required.

6. Placing of steel reinforcing which is incorporated in masonry.

7. Setting and incorporating into the masonry all bolts, anchors, metal attachments, nailing blocks and inserts, as furnished and located by others.

8. Building in of all door and window frames, vents, conduits, pipes, guards and inserts as furnished, set and braced by others.

9. Wire mesh backing for veneer over wood or metal construction.

Work Not Included:

1. Embedding anchoring devices in concrete.

2. Shoring and bracing.

3. Furnishing and fabricating of steel reinforcing.

4. Furnishing scratch coat backing for veneer.

Reinforced Grouted BRICK Masonry

Material

1. BUILDING (COMMON) BRICK: ASTM Designation C 62, Grade MW.
   Special size brick shall be _____
   Surface shall be (smooth) (Textured) (Cut Sides)

2. FACING BRICK: ASTM Designation C 126, Grade SW, Type FBS. Sizes shall be: (Standard, $2^3/16'' \times 3^1/2'' \times 7^1/2''$)
   (Norman, $2^3/16'' \times 3^1/2'' \times 11^1/2''$)
   (Continental, $2^3/8'' \times 3'' \times 11^1/2''$)

   Color and Texture shall be in accordance with approved samples.

3. FACING BRICK PAVERS: ASTM Designation C 216. Grade SW, 3000 psi, $2^3/16'' \times 3^1/2'' \times 7^1/2''$, $2^3/16'' \times 3^1/2'' \times 11^1/2''$, $1^1/2'' \times 3^1/2'' \times 11^1/2''$, $1^1/4'' \times 3^1/2'' \times 11^1/2''$, (plus other special sizes, check with local manufacturer.)

4. BUILDING BRICK PAVERS: $2^1/2'' \times 3^7/8'' \times 8^1/4''$ and $2^1/4'' \times 5^1/2'' \times 11^1/2''$.

5. BRICK BLOCK: $3'' \times 4^1/2'' \times 11^1/2''$, $3^1/2'' \times 7^1/2'' \times 11^1/2''$, $7^5/8'' \times 5^1/2'' \times 15^1/2''$, $7^5/8'' \times 7^1/2'' \times 15^1/2''$.

6. FIRE BRICK: ASTM Designation C 27: Low Duty, $2^1/2'' \times 4^1/2'' \times 9''$. Fire Brick splits, $1^1/4'' \times 4^1/2'' \times 9''$.

7. WATER: Taken from a supply distributed for domestic purposes, clean when used.

8. SAND: ASTM Designation C 144, except that not less than 5% shall pass the No. 100 sieve.

9. PEA GRAVEL: Graded with not more than 5% passing the No. 8 sieve and with 95% to 100% passing the $3/8''$ sieve.

10. PORTLAND CEMENT: ASTM Designation C 150, Type I or Type II, Low Alkali.

11. HYDRATED LIME: ASTM Designation C 207 Type S.

12. FLUE LINING: ASTM Designation C 315, made of Refractory Fire Clay. Concrete Flue Lining in accordance with ICBO RR No. 2602 and City of Los Angeles RR No. 23878.

13. REINFORCING STEEL: ASTM Designation A 615.

14. ADMIXTURE: No admixture shall be used in mortar or grout unless approved by the Architect or Engineer.

15. MORTAR COLOR: Color shall be an inert material as selected by the Architect.

16. SAMPLES OF BRICK: Before starting work, at least three (3) samples of each brick to be used shall be submitted for approval.

17. SAMPLE PANELS: Construct a sample panel approximately 3 feet high by 4 feet long of each type of brick to be used. Panel shall be one (1) tier thick, laid up with jointing as specified. If any panel is not satisfactory to the Architect, additional panels shall be constructed until approval is secured. The panels shall remain at job site until masonry work has been accepted.

Mortar and Grouting Mixes

All parts by volume measurement:

1. MORTAR: One(1) part portland cement and from one-quarter ($1/4$) to one-half ($1/2$) part lime putty or hydrated lime, and damp loose sand not less than two and one-quarter ($2 1/4$) and not more than three (3) times the sum of the volumes of cement and lime used.

2. GROUT: One (1) part portland cement and three (3) parts damp loose sand. Grout for spaces two inches (2") or more in width shall contain in addition, two (2) parts pea gravel (making a 1:3:2 mix). Sufficient water shall be added to provide pouring consistency without segregation.

3. SLUSH-GROUT: Sufficient water shall be added to mortar for placing with trowel.

Mixing Mortar and Grout

1. MIXING: All mortar and grout shall be mixed mechanically for not less than three (3) minutes. Ingredients shall be mixed carefully to provide a uniform mass.

2. RE-TEMPERING AND TIME LIMIT: Re-tempering on mortar boards shall be done by adding water within a basin formed in the mortar and the mortar re-worked into the water. Dashing or pouring water over mortar will not be permitted. Harsh, non-plastic mortar shall not be re-tempered or used.

Workmanship

1. WETTING OF BRICK: The amount of wetting will depend on the rate of absorption of the brick. When being laid, the bricks shall have sufficient suction to hold the mortar and to absorb the excess water from the grout.

2. LAYING BRICK: All brickwork shall be plumb, level and true to dimensions shown on plans.

3. BOND: Face bonding shall be as shown on the drawings, but if not shown, shall be half bond for standard size brick, third bond for oversize brick and quarter bond for modular brick.

4. STACK BOND: Brick shown to be laid in stack bond shall have the center lines of vertical joints plumb.

5. JOINTS: All bed and head joints shall be solidly filled with mortar or grout. The thickness of mortar joints shall be uniform and true to dimensions, consistent with brick dimensional tolerances. Type of joint shall be (tooled), (raked and tooled), (cut flush), (other as specified).

   Raked joints shall be raked out a maximum of three-eights of an inch ($3/8''$). Joints shall then be tooled with a flat jointing tool and finished by brushing a non-metallic brush.

6. POINTING, JOINTING AND TOOLING: When mortar has stiffened slightly all holes shall be filled with mortar. All jointing and tooling shall be done before mortar has set.

7. SACKING: Sack-rubbing or wiping finished masonry with rags will not be permitted.

8. PROTECTION: All adjoining work shall be protected from mortar droppings. The tops of all unfinished masonry shall be protected from rain. The masonry shall be protected by the masons while they are on job-site, and thereafter by others.

9. WETTING: Finished brickwork shall not be wetted except in hot weather when it may be fog sprayed, sufficient only to dampen surface.

Low Lift Grouted Masonry

Mortar in bed joints shall be held back one-fourth inch ($1/4''$) from edges of brick adjacent to grout spaces or shall be bevelled back and upward from grout space. Mortar droppings shall be kept out of grout spaces.

Head joints less than five-eighth inch ($5/8''$) thick shall be solidly filled with mortar. Head joints five-eighth inch ($5/8''$) or more may have mortar dams to retain the grout. Bed joints shall not be deeply furrowed. All bricks shall be shoved at least one-half inch ($1/2''$) into place.

In masonry which is more that two (2) tiers in thickness including pilasters and columns, the interior wall shall be of whole or half bricks placed into grout with not less than three-fourth inch ($3/4''$) of grout surrounding each brick or half brick. Except at the finish course, all grout shall be stopped one inch ($1''$) below the lower outer tiers.

When it is necessary to stop off a horizontal run of masonry, it must be done only by racking back in each course and stopping the grout two inches ($2''$) back of the rack. Toothing shall not be permitted.

One tier may be carried up eighteen inches ($18''$) before grouting, but the other tier should be laid up and grouted in lifts not exceeding eight inches ($8''$). Grout spaces must be immediately puddled with a stick or rod, not a trowel, to cause the grout to flow into all spaces between the brick. Vertical reinforcing must be held firmly in place. Horizontal reinforcing should be placed as work progresses. NOTE: Some codes may require that horizontal reinforcing shall be wired and held in place prior to start of bricklaying.

Reinforced Grouted BRICK Masonry

High Lift Grouted Construction

GENERAL: High-lift grouted construction must comply with the Standard Specifications for Reinforced Grouted Masonry except for the following: Grout Mixture, Mixing of Grout and Low-Lift Grouted Masonry Sections.

In addition, the following specifications shall apply:

1. FOUNDATION: The foundation surface which is to receive brickwork must be clean and damp, the laitance removed, by sandblasting if necessary, and the aggregate exposed.

2. LAYING BRICK: Mortar in all bed joints must be held back one-fourth inch (¼") from the edges of the brick adjacent to the grout space or should be bevelled back and upward from the grout space. The grout space must not be less than three inches (3") in width. The thickness of head and bed joints shall be as specified or shown. Head joints must be solidly filled with mortar as bricks are laid. Bed joints should not be deeply furrowed with the trowel. All bricks must be shoved at least one-half inch (½") into place. Where necessary to stop off a longitudinal run of masonry, it should be done only by racking back in each course. Toothing is not permitted.

3. WALL TIES: The two tiers must be bonded together with wall ties. Ties must be 9 gauge wire (ASTM A 82), in the form of rectangles four inches (4") wide and two inches (2") less than the thickness of the wall. No kinks, water drips or deformations are allowed. Ties will be laid not to exceed twenty-four inches (24") on center horizontally.

   Ties must be laid sixteen inches (16") on center vertically for running bond and twelve inches (12") on center vertically for stack bond. The ties should be placed in the bed joints of the two tiers simultaneously and aligned vertically so as not to interfere with vibrating.

4. CLEANOUTS: Cleanouts must be provided by leaving out every other unit in the bottom course on one (1) tier of the wall. During the work a high pressure jet stream of water should be used to remove mortar fins and all foreign matter from the grout space. The cleanout must be sealed after inspection and before grouting.

Masonry Institute of America

5. GROUT SPACE: The grout space must be not less than three inches (3") in width and should be poured solidly with grout. (NOTE: All construction under State of California Title 24 C.C.R. supervision requires a three and one-half inches (3½") minimum grout space).

6. BARRIERS: Vertical grout barriers or dams of solid masonry should be built across the grout space not more than thirty feet (30') apart. The barriers should be continuous the entire height of the wall.

7. REINFORCING: Horizontal reinforcing is required to be placed as work progresses. The vertical reinforcing is to be dropped into position after the completion of bricklaying prior to grouting. (NOTE: Some codes may require that vertical and horizontal reinforcing to be wired and held in place prior to start of bricklaying).

8. GROUT: Grout mix is to be one (1) part portland cement, three (3) parts sand and two (2) parts pea gravel, with the aggregates measured by volume in a damp, loose condition. The grout mix should contain a minimum of six (6) sacks of portland cement per yard. An admixture such as Suconem Grout Aid or equal must be used in the grout mix. One (1) pound of Grout Aid is to be added for each sack of cement with a maximum of six (6) pounds per yard.

9. GROUTING: It is recommended that no grout be pumped unless the brickwork has been allowed to set a minimum of three (3) days in hot weather or five (5) days in cold, damp weather. The grout mix is to be delivered in transit mix trucks. Water must be added so that the slump is near maximum without segregating. The grout should be pumped from the mixer into the grout space as rapidly as practical and discarded if not in place within one and one-half (1½) hours after water is first added to the batch. Grout must be placed in a continuous pour in lifts of approximately four (4) feet. It must be consolidated by mechanical vibration. The grouting of any section of wall between control barriers should be completed to the top of the wall in one (1) day.

10. BRACING: It is recommended that walls be adequately braced against wind and other forces during construction.

Reinforced Grouted BRICK Masonry

BRICK VENEER

1. BRICK VENEER: Brick for veneer shall be

_____

_____

_____

_____

2. GROUT SPACE: The veneer is to be laid not less than one inch (1″) clear of the backing. Grout space must be filled solidly with grout as each course is laid.

3. WALLS WITHOUT SHEATHING: The wood or metal studs, on sixteen inch (16″) centers, must be covered with a reinforced waterproof backing consisting of either "AQUA K-LATH or STEELTEX". On wood studs, it is recommended that the backing be nailed with galvanized barbed nails, at four inches (4″) on-centers (o.c.), penetrating a minimum of one and one-eighth inches (1 1/8″) into the stud. At the plates, 8d common nails shall be used at 8" o.c.

4. WOOD FRAME WALLS WITH SHEATHING: The sheathing should be covered with an asphalt impregnated building paper, lapped two inches (2″) at horizontal edges and six inches (6″) at vertical end joints. Veneer must be attached with corrosion-resistant metal clips not less than 1/16″ × 1″ nailed to the backing at studs and plates. The clips should extend into the mortar bed joints. They are to have continuous horizontal strand wire reinforcement of 9 gauge minimum, laid on the center line of bed joints of the veneer. Ties must be spaced not more than twelve inches (12″) apart vertically and twenty-four inches (24″) apart horizontally.

5. CONCRETE OR MASONRY CONSTRUCTION: It is recommended that veneer be attached with dovetail anchorages or other approved anchorages. Anchorages are to have a maximum spacing of twenty-four inches (24″) horizontally and twelve inches (12″) vertically. Anchors should have continuous nine (9) gauge wire reinforcement laid on the center line of bed joints.

Masonry Institute of America

CLAY BUILDING BRICK TEXTURES

The following names have been adopted by the brick manufacturers of Southern California as standards to aid in the design of better masonry. All textures are shown actual size.

SMOOTH

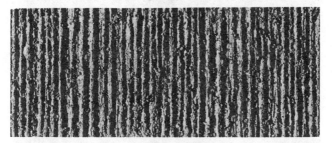
SCRATCH

WIRECUT

Reinforced Grouted BRICK Masonry

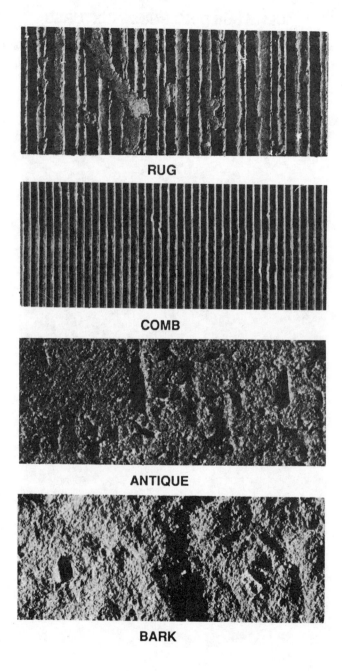

RUG

COMB

ANTIQUE

BARK

Masonry Institute of America

STONEFACE

INDEX

A

Absorption field test 9
Absorption of brick CCR 66
Absorption rate of 8
Acrylic-Resin 61
Acrylic-Resin coating 61
Admixtures 23
Anchorage of chimney 73
Anchorage of veneer 43, 44
Aqua K-Lath method 46
ASTM C 67 8

B

Bel air paver brick 2
Bituminous Compounds 62
Bond 3
    Quarter bond 3
    Third bond 3
Box fire 73
Bracing 49
Brebner 1
Brick 2
    Common building 2
    Modular building 2
    Oversize building 2
    Standard building 2
    Standard building 2
Brick Bel air paver 2
Brick building 6
Brick face 6
Brick inspection 77

Brick paving 2
Brick sizes 2
Brick specification 111
Brick split paver 2
Brick textures 119 – 121
Brick veneer 43, 118
Brick wetting 9
Bricklaying 34
Bricklaying CCR 66
Bricklaying inspection 77
Building brick 2, 6
Bulking sand 13

C

California Construction Requirements Title 24 65
Capping of sills 63
Capping of walls 63
Cement 66
Cement CCR 66
Cement masonry 17
Cement mortar 17
Cement plastic 15
Cement portland 15
Cement preblended 17
Cement-Water 61
Cement-Water coating 61
Cements ASTM C 150 151
Chimney 71
Chimney anchorage 73
Chimney parts 72
Cleaning 59
Coatings 61
Color mortar 23

Column ties 86
Columns 85
Common building brick 2
Common paver brick 2
Composition of coatings 61
Compression strength field test 57
Compression test grout 56
Considerations earthquake 69
Construction details 92 – 100
Coring 50
Coursing 4
Curing 50

D

Details construction 92 – 100
Dimensional tolerances 8
Dimensions 4

E

Earthquake considerations 69
Efflorescence 57
Elastomeric 62
Elastomeric coating 62

F

Face brick 2
    Imperial 2
    Modular 2
    Norman 2
    Oversize 2
    Regency 2
    Standard 2
Face Brick 2, 6
Field test Absorption 9

Field test compression strength 57
Field test of grout 55
Field test of mortar 55
Field test specimens of grout 54
Field test specimens of mortar 54
Field test specimens of mortar and grout 54
Fineness modulus sand 11
Fire box 73
Fire walls 62
Fireplace 71
Fireplace parts 72
Flush wall columns 85
Four-16≥ brick 2

G

Glossary of terms 101
Grading sand 11
Grout ASTM C 476 23
Grout CCR 66
Grout compression test 56
Grout field test 55
Grout field test specimens 54
Grout inspection 77
Grout mixing 114
Grout strength CCR 66
Grout test specimens 50
Grouting high lift 25, 31, 116
Grouting inspection 77
Grouting limitation 25
Grouting low lift 25, 26, 115
Grouting mixes 113
Grouting sequence 26
Guide specifications 111

H

Header 3
Hearth 73
High lift grouting 25, 116
High lift grouting procedure 31
History 1
Hollow brick 2
Hospitals Title 24 65
Hydrated lime 15

I

Imperial brick 2
Imperial face brick 2
Inspection 75
Inspection brick 77
Inspection bricklaying 77
Inspection grout 77
Inspection grouting 77
Inspection mortar 77
Inspection steel 77

K

Kanamori 1

L

Lap length 79, 80
Lap splices 78
Laying brick 34
Lime 15
Lime ASTM C 206 12
Lime ASTM C 5 12
Lime putty 15
Low lift grouting 25, 25, 115
Lump lime 15

M

Masonry cement 17
Masonry Core tests CCR 67
Materials storage 18
Method Aqua K-Lath 46
Method Steeltex 46
Mixing grout 114
Mixing mortar 114
Modular building brick 2
Modular face brick 2
Mortar ASTM C 270 19
Mortar CCR 66
Mortar cement 17
Mortar colors 23
Mortar field test 55
Mortar field test specimens 54
Mortar inspection 77
Mortar mixes 113
Mortar mixing 114
Mortar proportions 52
Mortar strength CCR 66
Mortar test specimens 50
Mortars water retention 20

N

Nomenclature 3
Norman face brick 2

O

Oil-base 62
Oversize face brick 2
Oversized building brick 2

P

Painting 59
Parapet 62
Parts chimney 72
Parts fireplace 72
Paver common 2
Paver split 2
Paving Brick 2
Planters 64
Plastering 65
Plastic cement 15
Poly-Vinyl Acetate 61
Poly-Vinyl Acetate coating 61
Porportions mortar 52
Portland cement 15
Preblended cement 17
Procedure high lift grouting 31
Processed lime 15
Protection of work 49
Public buildings Title 24 65

Q

Quater bond 3
Quicklime 15

R

Rate of absorption 8
Regency face brick 2
Reinforced grouted brick specification 111
Reinforcement 33
Requirements seismic 53
Retention water mortar 20
Rolock 3

Royal brick block 2
Running bond 3

S

Sand bulking 13
Sand CCR 66
Sand Fineness modulus 11
Sand grading 11
Sand voids 12
Sand, ASTM C 144 10
Scaffolding 49
Schools Title 24 65
Sealer 62
Seismic requirements 53
Sequence of grouting 26
Shear strength CCR 68
Shiner 3
Shoring 49
Silicone-Base 61
Silicone-Base coating 61
Sills capping 63
Sizes brick 2
Soldier 3
Spacing Tie 88
Spacing ties 87
Splice lap 78
Splice lap 78
Split paver brick 2
Splits 3
Stack Bond 3
Standard building brick size 2
Standard face brick 2
Steel inspection 77

Steel spacing CCR 66
Steel, ASTM A 615 18
Steeltex method 46
Storage of materials 18
Strechter 3
Synthetic-rubber 62

T

Test specimens grout 50
Test specimens mortar 50
Textures brick 119 – 121
Thames Tunnel 1
Third bond 3
Tie spacing 87, 88
Ties column 86
Tolerances dimensional 8
Two wythe grouting 26

V

Veneer anchorage 43, 44
Veneer brick 43, 118
Voids sand 12

W

Wall columns 85
Walls capping 63
Water 18
Water retention 20
Waterproofing 59
Wetting of brick 9
Work protection of 49
Workmanship 114